SAT II SUCCESS
Math IC & IIC

Mark N. Weinfeld
Lalit A. Ahuja
David Alan Miller

PETERSON'S
THOMSON LEARNING

Australia • Canada • Mexico • Singapore • Spain • United Kingdom • United States

About Peterson's

Founded in 1966, Peterson's, a division of Thomson Learning, is the nation's largest and most respected provider of lifelong learning online resources, software, reference guides, and books. The Education Supersite℠ at petersons.com—the Web's most heavily traveled education resource—has searchable databases and interactive tools for contacting U.S.-accredited institutions and programs. CollegeQuest® (CollegeQuest.com) offers a complete solution for every step of the college decision-making process. GradAdvantage™ (GradAdvantage.org), developed with Educational Testing Service, is the only electronic admissions service capable of sending official graduate test score reports with a candidate's online application. Peterson's serves more than 55 million education consumers annually.

Thomson Learning is among the world's leading providers of lifelong learning, serving the needs of individuals, learning institutions, and corporations with products and services for both traditional classrooms and for online learning. For more information about the products and services offered by Thomson Learning, please visit www.thomsonlearning.com. Headquartered in Stamford, Connecticut, with offices worldwide, Thomson Learning is part of The Thomson Corporation (www.thomson.com), a leading e-information and solutions company in the business, professional, and education marketplaces. The Corporation's common shares are listed on the Toronto and London stock exchanges.

Editorial Development: American BookWorks Corporation

Special thanks to Joan Marie Rosebush

For more information, contact Peterson's, 2000 Lenox Drive, Lawrenceville, NJ 08648; 800-338-3282; or find us on the World Wide Web at: www.petersons.com/about

COPYRIGHT © 2001 Peterson's, a division of Thomson Learning, Inc.
Thomson Learning™ is a trademark used herein under license.

Previous edition © 2000

ALL RIGHTS RESERVED. No part of this work covered by the copyright herein may be reproduced or used in any form or by any means—graphic, electronic, or mechanical, including photocopying, recording, taping, Web distribution, or information storage and retrieval systems—without the prior written permission of the publisher.

For permission to use material from this text or product, contact us by
Phone: 800-730-2214
Fax: 800-730-2215
Web: www.thomsonrights.com

ISBN 0-7689-0667-9

Printed in the United States of America

10 9 8 7 6 5 4 3 2 1   03 02 01

CONTENTS

Introduction 1
    About this Book 1
    About the Tests 2

SAT II Mathematics Study Plan 4
    The 10-Week Plan—2 Lessons per Week 4
    The 20-Week Plan—1 Lesson per Week 8
    The Panic Plan 8

DIAGNOSTIC TEST 1—MATHEMATICS LEVEL IC TEST 9
    Questions 12
    Mathematics Level IC Test Answers and Explanations 20

DIAGNOSTIC TEST 2—MATHEMATICS LEVEL IIC TEST 23
    Questions 26
    Mathematics Level IIC Test Answers and Explanations ... 34

MATHEMATICS IC AND IIC REVIEW 41
    Overview 42
    Arithmetic 42
    Properties of Numbers 43
    Systems of Measurements 79
    Signed Numbers 83
    Powers, Exponents, and Roots 88
    Algebra 95
    Plane Geometry 142
    Coordinate Geometry 178
    Functions and Their Graphs 185
    Arithmetic of Functions 188
    Trigonometry 195
    Trigonometric Functions of the General Angle 207
    Solving Triangles 224
    Scalars and Vectors 227
    Graphs of Trigonometric Functions 232
    Set Theory 238
    Probability 242
    Permutations and Combinations 248
    Statistics 252

CONTENTS

Exponents and Logarithms...............................	259
Logic...	266
Systems of Numbers.....................................	273
Complex Numbers.......................................	277
Sequences..	281

PRACTICE TEST 1—MATHEMATICS LEVEL IC TEST............... 289
Questions.. 292
Mathematics Level IC Test Answers and Explanations.... 307

PRACTICE TEST 2—MATHEMATICS LEVEL IC TEST............... 317
Questions.. 320
Mathematics Level IC Test Answers and Explanations.... 336

PRACTICE TEST 1—MATHEMATICS LEVEL IIC TEST.............. 343
Questions.. 346
Mathematics Level IIC Test Answers and Explanations... 362

PRACTICE TEST 2—MATHEMATICS LEVEL IIC TEST.............. 371
Questions.. 374
Mathematics Level IIC Test Answers and Explanations... 389

Answer Sheet... 395

INTRODUCTION

ABOUT THIS BOOK

Almost a quarter of a million students take SAT II Subject Tests every year. In the past, these tests were known as the College Board Achievement Tests. These tests are important for several reasons. Because many of the colleges require SAT II Subject Tests, these are important exams for you. The purpose of these tests is to measure and demonstrate your knowledge and/or skills in specific subjects and to test your ability to apply that knowledge to each particular examination. The better your score is, the better your application will look to the colleges of your choice.

If you're reading this book, it's likely that you are preparing for the SAT II Mathematics exam—either Level IC or Level IIC. We have tried to make this a workable book. In other words, the book is set up so that regardless of the level exam you're taking, you will be able to find the material necessary to study and to take those tests that are most applicable to your level.

As a further enhancement to your ability to prepare for this exam, we have prepared in-depth mathematics review material and highlighted those areas that are required primarily for Level IIC, so that those studying for the Level IC test can focus on only those areas that are appropriate.

Divided into sections, the book begins with two diagnostic exams. There is one each for Level IC and Level IIC. Take these exams (and all of the tests) under simulated exam conditions, if you can. Find a quiet place in which to work, set up a clock, and take the test without stopping. When you are finished, take a break and then go back and check your answers. Always reread those questions you got wrong, since sometimes your errors came from merely misreading the question. Again, double-check your answers, and if they're still not clear, read the appropriate section in the review material.

Once you've completed your diagnostic test(s), it's time to move on to the review section. Study the material carefully, but feel free to skim the portion of the review section that is easy for you.

Then, take the actual practice tests. These simulated exams are designed to give you a broad spectrum of question types that are similar to those you will find on the actual SAT II Mathematics tests. We suggest that, regardless of the level exam you are planning to take, it would be extremely helpful to *take all of the tests in the book*. If you are taking Level II, taking the lower-level test will give you that much more practice for the exam. And if you are taking Level I, it would be helpful to test your skills and stretch your thinking to give you a stronger grounding for the Level I exam.

As you complete each exam, take some time to review your answers. We think you'll find a marked improvement from taking the diagnostic tests to completing all of the full-length practice tests. Always take the time to check the review section for clarification, and if you still don't understand the material, go to your teacher for help.

About the Tests

Each of the SAT II Mathematics Tests (Level IC and Level IIC) is similar in format. They both contain 50 multiple-choice questions. In addition, the current versions of the tests includes a background questionnaire on the first page. The College Board uses this information for statistical purposes; your answers will not affect your scores in any way.

Contents

The Level IC examination tests material that covers the following topics:
- Algebra (30%)
- Geometry (plane Euclidean, three-dimensional, coordinate) (38%)
- Basic Trigonometry (8%)
- Algebraic functions (12%)
- Elementary statistics (probability, counting problems, data interpretation, mean, median, and mode) (6%)
- Miscellaneous topics (logic, number theory, arithmetic, geometric sequences) (6%)

The Level IIC examination tests material that covers the following topics:
- Algebra (18%)
- Geometry (coordinate and three-dimensional) (20%)
- Trigonometry (20%)
- Functions (24%)
- Statistics (probability, permutations, combinations) (6%)
- Miscellaneous topics (logic and proof, number theory, sequences, limits) (12%)

You can quickly see where your focus should be for each level, depending upon the percentage of questions that are on the tests.

INTRODUCTION

Calculators

You may use almost any scientific or graphic calculator, and it is estimated that you will find it useful, and often necessary, for about 40 percent of the questions on the Level IC exam and for about 60 percent on the Level IIC exam. You should, therefore, be very familiar with the operation of the calculator you plan to bring to the exam.

Scoring

While it's not imperative that you completely understand how the test is scored, since the process shouldn't deter you from trying to do your best, you are probably aware that the scores are reported on the 200–800 range.

Each question answered correctly receives one point. You lose a fraction of a point for each incorrect answer. However, you do not lose points if you don't answer a question. (If you skip any answers, make sure that the next question you answer is filled in on your answer page in the correct space.) Thus, it make sense to guess at those questions that you don't know, and of course, as with most multiple-choice questions, you should use the process of elimination to increase the odds of guessing correctly. The more choices you eliminate, the better your odds are for choosing the correct answer.

Taking the Test

Since you have 60 minutes in which to complete the exam, it is important that you pace yourself. One very important item to remember is to be thoroughly familiar with the directions for the tests so that you don't waste time trying to understand them once you've opened your test booklet.

Work through the easy questions first. The faster you can complete those questions, the more time you'll have for those that are more difficult. You may use your test book for scratch paper, but keep your answer sheet clean since they are machine-readable, and any stray marks might be construed as an answer.

In order to take either of these tests, you will have had two to three years of college preparatory math, so you should be well-prepared to take either one of these exams. You probably won't need to worry about whether or not you know the material, especially if you carefully use this book.

Good luck!

SAT II MATHEMATICS STUDY PLAN

Everyone needs a plan, especially when you're preparing for the SAT II Mathematics exam. The first thing you need to do is to figure out how much time you have before the exam. If you have enough time to prepare completely, it'll make your studying easier. If, however, you're somewhat short on time, this plan will be even more valuable for you. We offer you these different study plans to help maximize your time and studying. The first is a 10-Week Plan, which involves concentrated studying and a focus on the sample test results. The second is the more leisurely 20-Week Plan, one that's favored by schools. Finally, if time is running short, you should use the Panic Plan. We don't want you to panic, however—this plan is supposed to help you conquer that panic and help you organize your studying so that you can get the most out of your review work and still be as prepared as possible.

These plans are supposed to be flexible and are only suggestions. Feel free to modify them to suit your needs and your own study habits. But start immediately. The more you study and review the questions, the better your results will be.

THE 10-WEEK PLAN—2 LESSONS PER WEEK

Week 1

Lesson 1: **Diagnostic Test #1**
Take the Diagnostic SAT II Mathematics—Level IC Test in its entirety. It is designed to help you determine what you need to know and where to focus your studying. Take this test under simulated test conditions, in a quiet room, and keep track of the time it takes to complete the test.

Diagnostic Answers
Once you have completed the test, check all of your answers and read through the explanations. This may take quite a bit of time, as will all of the tests, but it will enable you to select those subject areas on which you should focus.

Lesson 2: **Diagnostic Test #2**
Take the Diagnostic SAT II Mathematics—Level IIC Test under the same conditions as you took the Level IC exam. It is designed to help you determine what you need to know at this level and where to focus your studying. Even if you are planning to take only the Level IC exam, it will be a good test of what you know if you also take the Level IIC exam. It will force you to stretch your thinking—and you might even surprise yourself by how much you understand.

Diagnostic Answers
Once you have completed the test, check all of your answers and read through the explanations.

INTRODUCTION

Week 2

Lesson 1: **Review of Basic Arithmetic**
Take your time to read through this first section. Note that the style of the review material is in an outline format. It should be similar to your classroom notes. Underline or use a marker to highlight those areas that are unclear to you.

Lesson 2: **Review of Algebra**
This section presents the basics of solving equations, word problems with one unknown, literal equations, linear inequalities, ratio and proportion, variation, and means and medians. Study the material carefully, and answer the review questions at the end of each section.

Week 3

Lesson 1: **Review of Geometry**
As you continue your studying, try to work in a quiet room, uninterrupted by others in you household or the TV, radio, or any outside noises.

Lesson 2: **Functions and Their Graphs, and Right Triangle Trigonometry**
Again, read through this chapter, mark whatever is unclear, and go back and reread the material, if necessary. You may find it easier to break this section into smaller ones and do a little at a time.

Week 4

Lesson 1: **Trigonometric Functions of General Angles: Part One**
You're about half way through the content chapters of this book now, so continue reading and taking notes. It might be smart to work on half the chapter in the morning and the other half in the afternoon. This section contains trigonometric definition of angles, degrees and radians, trigonometric functions, and reference angles.

Lesson 2: **Trigonometric Functions of General Angles: Part Two**
Read through this chapter, mark whatever is unclear, and then go back and reread the material, if necessary. You can always ask your teacher for additional information if you're having difficulty. This section presents trigonometric equations, identities, solving triangles, and graphs of trigonometric functions.

Week 5

Lesson 1: **Set Theory, Probability, and Statistics**
There are only two more sections after this one. By now, you should have a strong understanding of the material.

Lesson 2: **Exponents and Logarithms**
All that's left is one more chapter that will help you improve your scores on the actual exam. Again, make sure you mark whatever you don't understand. You'll have time to review later on or to ask you teacher for help.

Week 6

Lesson 1: **Logic, Systems of Numbers, and Sequences and Series**
This is the end of the review chapters. How did you do? Did you correctly answer most of the questions at the end of each section?

Lesson 2: **Review**
This is the time to take a breather and go back and look over the content chapters of the book in order to find anything that might have slipped by. Go back to the questions that gave you problems. Do you understand them now? Try to answer some questions in earlier chapters. You have some time before taking the practice tests.

Week 7

Lesson 1: **SAT II Level IC Mathematics Practice Test #1**
Regardless of the level test you are planning to take, you should take this test (and all of the others that follow) and answer all of the questions you can, and then guess at those you don't know. Circle those questions that you guessed at so that you can zero in on those specific answers and so that you don't delude yourself into thinking that you really knew those answers in the first place.

Lesson 2: **SAT II Level IC Mathematics Practice Test #1 Answers**
Check all of your answers, and review the material.

Week 8

Lesson 1: **SAT II Level IC Mathematics Practice Test #2**
Take this test and answer all of the questions you can. By now, you will have noticed your improvement from when you took the diagnostic test to now.

Lesson 2: **SAT II Level IC Mathematics Practice Test #2 Answers**
Check all of your answers, and review the material. If you are taking the Level IC exam, you should now go back to check anything on these last two tests that you did not understand.

Week 9

Lesson 1: **SAT II Level IIC Mathematics Test #3**
We suggest that you take this level test even if you're planning to take the Level IC exam, since it will give you additional practice on a wide range of material and even include much of the material that appears on the Level IC test.

Lesson 2: **SAT II Level IIC Mathematics Practice Test #3 Answers**
Check all of your answers, and review the incorrect answers. Before moving on, you should have a clear understanding of why you answered questions incorrectly.

Week 10

Lesson 1: **SAT II Level IIC Mathematics Test #4**
This is the final test. By now, you will have answered hundreds of multiple-choice questions throughout this book, including the diagnostic tests, practice tests, and review questions. Although there may be some new material on the actual test that you haven't encountered here, you should be very well prepared.

Lesson 2: **SAT II Level IIC Mathematics Practice Test #4 Answers**
Check all of your answers, and review the material. Then rest!

THE 20-WEEK PLAN—1 LESSON PER WEEK

The 20-Week Plan is for those of you who have the extra time, and it will enable you to better utilize your study time. You can now spread out your plan into one lesson per week. This plan is ideal because you are not under any pressure, and you can take more time to review the material in each of the chapters. You will also have enough time to double-check the answers to those questions that might have given you problems. Keep in mind that the basis for all test success is practice, practice, practice.

THE PANIC PLAN

We really don't want you to panic, but this plan is for those of you who do not have the luxury of extra time to prepare for the SAT II Mathematics test. Perhaps, however, we can offer you a few helpful hints to get you through this period.

Read through the official SAT II Mathematics bulletin and this *SAT II Mathematics Success* book, and memorize the directions. One way of saving time on this, or any, test is to be familiar with the directions in order to maximize the time you have to work on the questions. On this test, they're pretty simple.

Read the introduction to this book. It will be helpful in preparing for the test, and it gives you an understanding of what you can expect on the exam and how much time you will have to complete both sections of the test.

Take the diagnostic tests as well as the practice tests. Although we recommend that you take every test in the book, regardless of the actual level test you plan to take, if you don't have the time, just focus on that level that will be appropriate for you. If you have enough time, go back and check the questions and answers within the review sections. Focus whatever time you have left on those specific areas of the test that gave you the most difficulty when you took the practice tests. Whatever time you have before the exam, keep in mind that the more you practice, the better you will do on the final exam.

Diagnostic Test 1

MATHEMATICS LEVEL IC TEST

MATHEMATICS LEVEL IC TEST

While you have taken many standardized tests and know to blacken completely the ovals on the answer sheets and to erase completely any errors, the instructions for the SAT II Mathematics IC exam differs in three important ways from the directions for other standardized tests you have taken. You need to indicate on the answer key which test you are taking.

The instructions on the answer sheet will tell you to fill out the top portion of the answer sheet exactly as shown.

1. Print MATHEMATICS LEVEL IC on the line to the right under the words *Subject Test (print)*.

2. In the shaded box labeled *Test Code* fill in four ovals:

   —Fill in oval 3 in the row labeled V.
   —Fill in oval 2 in the row labeled W.
   —Fill in oval 5 in the row labeled X.
   —Fill in oval A in the row labeled Y.
   —Leave the ovals in row Q blank.

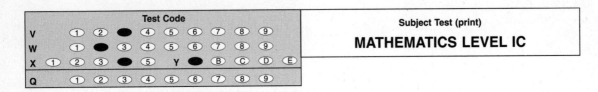

When everyone has completed filling in this portion of the answer sheet, the supervisor will tell you to turn the page and begin the Mathematics Level IC examination. The answer sheet has 100 numbered ovals on the sheet, but there are only 60 multiple-choice questions in the test, so be sure to use only ovals 1 to 50 to record your answers.

MATHEMATICS LEVEL IC TEST

REFERENCE INFORMATION

The following information is for your reference in answering some of the questions in this test.

Volume of a right circular cone with radius r and height h:
$$V = \frac{1}{3}\pi r^2 h$$

Lateral area of a right circular cone with circumference of the base c and slant height l:
$$S = \frac{1}{2}cl$$

Surface area of a sphere with radius r:
$$S = 4\pi r^2$$

Volume of a pyramid with base area B and height h:
$$V = \frac{1}{3}Bh$$

For each of the following problems, identify the BEST answer of the choices given. If the exact numerical value is not one of the choices, select the answer that is closest to this value. Then fill in the corresponding oval on the answer sheet.

NOTES

1. You will need a calculator to answer some (but not all) of the questions. You must decide whether or not to use a calculator for each question. You must use at least a scientific calculator; you are permitted to use graphing or programmable calculators.

2. Degree measure is the only angle measure used on this test. Be sure your calculator is set in degree mode.

3. The figures that accompany questions on the test are designed to give you information that is useful in solving problems. They are drawn as accurately as possible EXCEPT when stated that a figure is not drawn to scale. Unless otherwise indicated, all figures lie in planes.

4. The domain of any function f is assumed to be the set of all real numbers x for which $f(x)$ is a real number, unless otherwise specified.

DIAGNOSTIC TEST 1

MATHEMATICS LEVEL IC TEST

USE THIS SPACE FOR SCRATCH WORK

1. If $x = 2$, then $(x - 1)(2x - 3) =$

   (A) 1
   (B) 3
   (C) -1
   (D) 0
   (E) -3

2. The side of the inside square is x; the side of the outside square is $x + 2$. How much larger is the area of the outside square than the area of the inside square?

   (A) 4
   (B) $4x$
   (C) $4x + 4$
   (D) $4x + 1$
   (E) $x + 4$

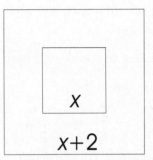

3. Given the figure to the right, what is the area in terms of a and b of the large square?

   (A) $a^2 + b^2$
   (B) $(a + b)^2$
   (C) $(a - b)^2$
   (D) c^2
   (E) $(a - 2b)^2 + b^2$

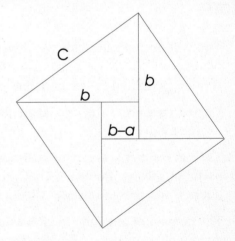

12

Peterson's SAT II Success:
Mathematics IC and IIC

MATHEMATICS LEVEL IC TEST—Continued

4. What is the sine of 30 degrees?

    (A) $\dfrac{1}{2}$

    (B) $\dfrac{3}{2}$

    (C) $\dfrac{\sqrt{2}}{2}$

    (D) $\dfrac{\sqrt{3}}{2}$

    (E) 2

5. A gumball machine contains 40 blue gum balls, 20 red gumballs, 15 green gumballs, and 25 purple gumballs. What is the probability that a person gets a red gumball?

    (A) .10
    (B) .20
    (C) .25
    (D) .15
    (E) .30

6. Find x if $\dfrac{x}{4} + \dfrac{(x+1)}{3} = \dfrac{3}{2}$.

    (A) $\dfrac{17}{2}$

    (B) $\dfrac{15}{7}$

    (C) 2

    (D) 1

    (E) -2

MATHEMATICS LEVEL IC TEST—Continued

7. At what point(s) does $f(x) = x^2 + 3x + 2$ intersect the x-axis?

   (A) (2, 0) and (1, 0)
   (B) (0, 2) and (0, 1)
   (C) (0, −1) and (0, −2)
   (D) (−1, 0) and (−2, 0)
   (E) (−1, 0) and (2, 0)

8. Sequential arrangements of blocks are formed according to a pattern. Each arrangement after the first one is generated by adding a row of squares to the bottom of the previous arrangement, as shown in the figure to the right. If this pattern continues, which of the following gives the number of squares in the nth arrangement?

   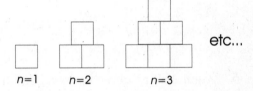

   (A) $2n^2$
   (B) $2(2n - 1)$
   (C) $\dfrac{n(n + 1)}{2}$
   (D) $n(n + 1)$
   (E) $n^2 + 1$

9. If the distance between the two points $P(a, 3)$ and $Q(4, 6)$ is 5, then find a.

   (A) 0
   (B) −4
   (C) 8
   (D) −4 and 0
   (E) 0 and 8

MATHEMATICS LEVEL IC TEST—Continued

USE THIS SPACE FOR SCRATCH WORK

10. Line m has a negative slope and a negative y-intercept. Line n is parallel to line m and has a positive y-intercept. The x-intercept of n must be

   (A) negative and greater than the x-intercept of m
   (B) negative and less than the x-intercept of m
   (C) zero
   (D) positive and greater than the x-intercept of m
   (E) positive and less than the x-intercept of m

11. Given the list of numbers {1, 6, 3, 9, 16, 11, 2, 9, 5, 7, 12, 13, 8}, what is the median?

   (A) 7
   (B) 8
   (C) 9
   (D) 11
   (E) 6

12. If $f(x) = x + 1$ and $g(x) = x + 1$, then find $f(g(x))$.

   (A) $(x + 1)^2 + 1$
   (B) $x^2 + x + 2$
   (C) $x + 2$
   (D) $x^3 + x^2 + x + 1$
   (E) $x^2 + 2x + 2$

13. The equation $x^2 + 9 = 2y^2$ is an example of which of the following curves?

   (A) hyperbola
   (B) circle
   (C) ellipse
   (D) parabola
   (E) line

GO ON TO THE NEXT PAGE

MATHEMATICS LEVEL IC TEST—Continued

14. If $x + 1 < 0$ then $|x|$ is

    (A) 0
    (B) $x + 1$
    (C) $-x$
    (D) $-x - 1$
    (E) x

15. The angle of the sun above the horizon is 27.5 degrees. Find the length of the shadow of a person who is 4.75 feet tall.

    (A) 4.75
    (B) 2.47
    (C) 4.65
    (D) 9.12
    (E) 4.86

16. Define the x-coordinate of the midpoint between $(-1, 1)$ and $(4, 1)$ as m and define the y-coordinate of the midpoint between $(4, 1)$ and $(4, 3)$ as n. What are the x- and y-coordinates of (m, n)?

    (A) $\left(\frac{5}{2}, \frac{3}{2}\right)$
    (B) $\left(\frac{5}{2}, 2\right)$
    (C) $\left(\frac{3}{2}, \frac{3}{2}\right)$
    (D) $\left(\frac{3}{2}, 2\right)$
    (E) $(4, 1)$

17. What is the slope of the line that is perpendicular to $y - 2x = 1$?

    (A) 2
    (B) −2
    (C) $\frac{1}{2}$
    (D) $-\frac{1}{2}$
    (E) 1

18. Trees are to be planted inside a circular tree orchard so that there are 5 trees per square meter. The circumference of the tree orchard is 30 meters. If trees are available only in allotments of packages of 6, how many allotments will the caretaker need to purchase?

    (A) 59
    (B) 117
    (C) 58
    (D) 60
    (E) 118

19. Place the following list of numbers with the given labels in order of largest to smallest.

    $F = 10^{10^{10}}$, $G = 10^{10}10^{10}$, $H = \dfrac{10^{100}}{10^{10}}$, $I = 10$

    (A) F, G, H, I
    (B) G, F, H, I
    (C) F, H, G, I
    (D) H, F, G, I
    (E) G, H, F, I

MATHEMATICS LEVEL IC TEST—Continued

20. The statement "y varies inversely with the square of x" means

    (A) $y = kx$
    (B) $y = \dfrac{k}{x}$
    (C) $y = k$
    (D) $y = \dfrac{k}{x^2}$
    (E) $y = kx^2$

21. Suppose one takes a tetrahedron *ABCD* with base *ACD* as shown to the right and combines it with another tetrahedron by connecting their corresponding bases. How many distinct edges does the resulting object have?

    (A) 9
    (B) 12
    (C) 10
    (D) 11
    (E) 8

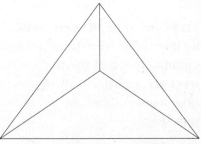

Looking down on the Tetrahedron.
Base is on the bottom.

22. The area of the parallelogram *ABCD* is

    (A) 20
    (B) 10
    (C) $5\sqrt{3}$
    (D) $10\sqrt{3}$
    (E) 18

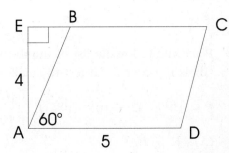

MATHEMATICS LEVEL IC TEST—Continued

USE THIS SPACE FOR SCRATCH WORK

23. If $f(x) = 3x - 1$ and if f^{-1} is the inverse function of f, what is $f^{-1}(5)$?

    (A) 18
    (B) $\frac{4}{3}$
    (C) $\frac{5}{3}$
    (D) 2
    (E) 0

24. If $(x - 1)^{x^2} = 1$ and x is not equal to 1, then all the possible values of x are

    (A) 0
    (B) 2
    (C) −2
    (D) 0 and 2
    (E) none of the above

25. What is the sum of the first 40 even positive integers?

    (A) 1,600
    (B) 1,560
    (C) 820
    (D) 1,640
    (E) 400

STOP If you finish before time is called, you may check your work on this section only. Do not turn to any other section in the test.

DIAGNOSTIC TEST 1

Quick Score Answers

1. A	6. C	11. B	16. D	21. A
2. C	7. D	12. C	17. D	22. A
3. A	8. C	13. A	18. D	23. D
4. A	9. E	14. C	19. C	24. D
5. B	10. D	15. D	20. D	25. D

ANSWERS AND EXPLANATIONS

1. **The correct answer is (A).** If $x = 2$ then $(x - 1)(2x - 3) = (2 - 1)(2(2) - 3) = 1(1) = 1$.

2. **The correct answer is (C).** The area of the larger square is $(x + 2)^2$, and the area of the smaller square is x^2. Hence, the area of the larger square is $(x + 2)^2 - x^2 = 4x + 4$ larger than the area of the smaller square.

3. **The correct answer is (A).** Use the Pythagorean theorem on one of the triangles to get $c^2 = a^2 + b^2$.

4. **The correct answer is (A).** Use a 30-60-90 special triangle to get $\sin(30) = \frac{1}{2}$.

5. **The correct answer is (B).** The probability that a person gets a red gumball is just the number of red gumballs divided by the total number of gumballs. That is, $\frac{20}{100}$.

6. **The correct answer is (C).** If $\frac{x}{4} + \frac{(x + 1)}{3} = \frac{3}{2}$ then add the two algebraic fractions by first getting a common denominator of 12 and then adding the two fractions. The result of this is $\frac{7x + 4}{12} = \frac{3}{2}$. Now multiply both sides of the equation by 12 to clear the denominators. The result of this is $7x + 4 = 18$, and now solving for x, we get the desired result.

7. **The correct answer is (D).** Set $f(x) = 0$ and solve by factoring and using the zero-factor property. That is, $f(x) = (x + 2)(x + 1) = 0$ then $x + 2 = 0$ and $x + 1 = 0$, which says the x-intercepts are $(-2, 0)$ and $(-1, 0)$.

8. **The correct answer is (C).** Since we are adding the numbers from 1 up to the number that n equals, we use the formula $\frac{n(n + 1)}{2}$.

9. **The correct answer is (E).** The distance between the two points P and Q is defined to be 5. That is, $\sqrt{(a - 4)^2 + (3 - 6)^2} = 5$. To find the answer, we need to solve this equation for a. So, after squaring both sides and simplifying, we get the equation $(a - 4)^2 = 16$. So, $a = 0$ and 8.

ANSWERS AND EXPLANATIONS

10. **The correct answer is (D).** If line m has a negative slope and a negative y-intercept, it has a negative x-intercept. Since line n is parallel to line m with a positive y-intercept we know that line n will have a positive x-intercept. So, the answer follows.

11. **The correct answer is (B).** Given the list, first write the list in increasing order from the smallest number to the largest number and then find the number that has an equal number of numbers on each side. Thus, we find that 8 is the number that has 6 numbers below it and 6 above it.

12. **The correct answer is (C).** Evaluate f at $g(x)$ to get $f(g(x)) = g(x) + 1$, and now substitute $g(x)$ to get the answer.

13. **The correct answer is (A).** Just rewrite the equation to $2y^2 - x^2 = 9$ to see that it is a hyperbola.

14. **The correct answer is (C).** If $x + 1 < 0$ then $x < -1$. So, the absolute value of a negative number is defined to be the same number but positive. So, $|x| = -x$.

15. **The correct answer is (D).** Given the information, we can draw the following triangle:

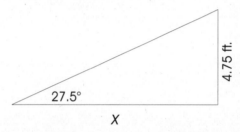

The answer can be found by looking at the tangent of 27.5 and solving for x.

16. **The correct answer is (D).** Calculate the average of the x coordinates of -1 and 4 (define as m) and also calculate the average of the y coordinates 1 and 3 (define as n). So, (m, n) is given by choice (D).

17. **The correct answer is (D).** Write in slope-intercept form to get the slope of the given line to be $2 = m_1$. Thus, the line that is perpendicular to the given line has slope $\frac{-1}{m_1} = -\frac{1}{2}$.

18. **The correct answer is (D).** Circumference is given by the formula $C = 2\pi r$. So, using the given information, we can calculate that $r = \frac{30}{(2\pi)}$. Now plug r into $A = \pi r^2$ to get the area in meters squared. We know that there are 5 trees per square meter and there are 6 trees in a package, so multiply the last result by 5 and divide by 6. That is,

$$\frac{5A}{6} \approx \frac{5}{6}(71.62) \approx 59.68.$$

Therefore, we would need to buy 60 packages.

19. **The correct answer is (C).** By exponent rules, $F = 10^{100}$, $G = 10^{20}$, $H = 10^{90}$, and $I = 10$. So, the order of smallest to largest is choice (C).

20. **The correct answer is (D).** Y varies inversely with the square of x, which means $y = \dfrac{k}{x^2}$.

21. **The correct answer is (A).** One tetrahedron has 6 edges, and if we place two together along the bases of the two tetrahedrons, they share 3 common edges for a total of $12 - 3 = 9$ edges.

22. **The correct answer is (A).** The area of the parallelogram is just $4(5) = 20$.

23. **The correct answer is (D).** $f^{-1}(5)$ implies that $f(x) = 5$, which says $3x - 1 = 5$ or $x = 2$.

24. **The correct answer is (D).** For $(x - 1)x^2 = 1$, either $x - 1 = \pm 1$ or $x^2 = 0$ because the only values that can be raised to a non-zero power to yield 1 are $+1$ and -1. $x - 1 = 1 \Rightarrow x = 2$; $x - 1 = -1 \Rightarrow x = 0$; $x^2 = 0 \Rightarrow x = 0$; hence $x = 2$ or $x = 0$.

25. **The correct answer is (D).** $2 + 4 + 6 + \ldots + 80 = 2[1 + 2 + 3 + \ldots + 40]$. The sum of the first n positive integers is $\dfrac{n \cdot (n+1)}{2}$, so the total is $2\left[\dfrac{40 \cdot 41}{2}\right] = 1{,}640$.

Diagnostic Test 2

MATHEMATICS LEVEL IIC TEST

MATHEMATICS LEVEL IIC TEST

While you have taken many standardized tests and know to blacken completely the ovals on the answer sheets and to erase completely any errors, the instructions for the SAT II Mathematics IIC exam differs in three important ways from the directions for other standardized tests you have taken. You need to indicate on the answer key which test you are taking.

The instructions on the answer sheet will tell you to fill out the top portion of the answer sheet exactly as shown.

1. Print MATHEMATICS LEVEL IC on the line to the right under the words *Subject Test (print)*.

2. In the shaded box labeled *Test Code* fill in four ovals:

   —Fill in oval 3 in the row labeled V.
   —Fill in oval 2 in the row labeled W.
   —Fill in oval 5 in the row labeled X.
   —Fill in oval A in the row labeled Y.
   —Leave the ovals in row Q blank.

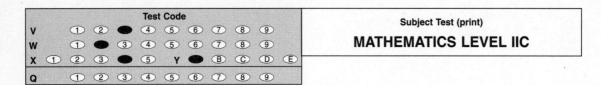

When everyone has completed filling in this portion of the answer sheet, the supervisor will tell you to turn the page and begin the Mathematics Level IIC examination. The answer sheet has 100 numbered ovals on the sheet, but there are only 60 multiple-choice questions in the test, so be sure to use only ovals 1 to 50 to record your answers.

MATHEMATICS LEVEL IIC TEST

REFERENCE INFORMATION

The following information is for your reference in answering some of the questions in this test.

Volume of a right circular cone with radius r and height h:
$$V = \frac{1}{3}\pi r^2 h$$

Lateral area of a right circular cone with circumference of the base c and slant height l:
$$S = \frac{1}{2}cl$$

Surface area of a sphere with radius r:
$$S = 4\pi r^2$$

Volume of a pyramid with base area B and height h:
$$V = \frac{1}{3}Bh$$

For each of the following problems, identify the BEST answer of the choices given. If the exact numerical value is not one of the choices, select the answer that is closest to this value. Then fill in the corresponding oval on the answer sheet.

NOTES

1. You will need a calculator to answer some (but not all) of the questions. You must decide whether or not to use a calculator for each question. You must use at least a scientific calculator; you are permitted to use graphing or programmable calculators.

2. Degree measure is the only angle measure used on this test. Be sure your calculator is set in degree mode.

3. The figures that accompany questions on the test are designed to give you information that is useful in solving problems. They are drawn as accurately as possible EXCEPT when stated that a figure is not drawn to scale. Unless otherwise indicated, all figures lie in planes.

4. The domain of any fuction f is assumed to be the set of all real numbers x for which $f(x)$ is a real number, unless otherwise specified.

DIAGNOSTIC TEST 2

MATHEMATICS LEVEL IIC TEST

USE THIS SPACE FOR SCRATCH WORK

1. $b\left(\dfrac{3a}{b} + \dfrac{2}{c}\right) =$

   (A) $\dfrac{3ab + 2}{bc}$

   (B) $\dfrac{3ac + 2b}{c}$

   (C) $\dfrac{3a + 2b}{bc}$

   (D) $\dfrac{3a + 2}{c}$

   (E) $\dfrac{3ab + 2b}{b + c}$

2. If $2^x \cdot 2^y = 3^z$, $\dfrac{x+y}{z} =$

   (A) 1.330
   (B) 1.750
   (C) 0.750
   (D) 1.585
   (E) 1.425

3. $(a + 2i)(b - i) =$

   (A) $a + b - i$
   (B) $ab + 2$
   (C) $ab + (2b - a)i + 2$
   (D) $ab - 2$
   (E) $ab + (2b - a)i - 2$

MATHEMATICS LEVEL IIC TEST—Continued

USE THIS SPACE FOR SCRATCH WORK

4. In a quadrilateral with angles A, B, C, and D, if A = B and C is one half of A while D exceeds A by 10°, what are the values of A, B, C, and D, respectively?

   (A) 100, 100, 150, 10
   (B) 75, 75, 150, 60
   (C) 100, 100, 50, 110
   (D) 90, 90, 45, 100
   (E) 120, 120, 60, 130

5. If an item is purchased at $150 and sold at $165, what percent of the original cost is the profit?

   (A) 110 percent
   (B) 89 percent
   (C) 100 percent
   (D) 9 percent
   (E) 10 percent

6. $f(x) = \dfrac{x^2 - 1}{x + 3}$ and $g(f(5)) = 21$, then which of the following could be $g(t) =$

   (A) $3t^2 + 2t + 1$
   (B) $t^2 - t + 1$
   (C) $t^2 + 3t - 1$
   (D) $t^2 + 3t + 3$
   (E) $\dfrac{t^2 + 3}{-1}$

GO ON TO THE NEXT PAGE

7. Consider a pyramid with a square base side of 6 inches and with a height of 12 inches, as shown on the right. If we cut off the top of the pyramid parallel to the base 3 inches from the tip, what is the volume of the remaining solid?

   (A) 141.75
   (B) 140
   (C) 135.48
   (D) 144
   (E) 130

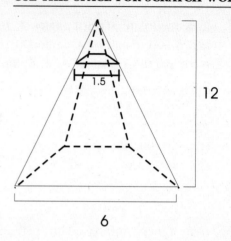

8. If $f(x) = ax^2 + bx + c$ and $f(0) = 1$ and $f(-1) = 3$, $(a - b) =$

   (A) 3
   (B) 1
   (C) 2
   (D) 0
   (E) -1

9. The domain of the function $f(x) = \dfrac{x - 4}{\sqrt{1 - 2x}}$ is

   (A) $x \neq \dfrac{1}{2}$
   (B) $x \geq 4$
   (C) $x \neq 4$
   (D) $x < \dfrac{1}{2}$
   (E) $x > 0$

10. A bug travels all the way around a circular path in 30 minutes traveling at 62.84 inches per hour. What is the radius of the circular path?

   (A) 10 inches
   (B) 3.142 inches
   (C) 7.5 inches
   (D) 12.5 inches
   (E) 5 inches

11. What is the length of an arc of a circle with a radius of 5 if it subtends an angle of 60° at the center?

   (A) 3.14
   (B) 5.24
   (C) 10.48
   (D) 2.62
   (E) 4.85

12. If we have a function
   $f(x) = x^2 - 6x, \quad x \leq 2$
   $ = 2x - 1, \quad x > 2,$
   $f(3) =$

   (A) -9
   (B) 5
   (C) 8
   (D) 6
   (E) Not defined

MATHEMATICS LEVEL IIC TEST—Continued

USE THIS SPACE FOR SCRATCH WORK

13. The equation of the tangent to a circle at point (6, 6) if the circle has its center at point (3, 3), would be

   (A) $x + y - 12 = 0$
   (B) $x = y$
   (C) $x + y = 0$
   (D) $x + y + 12 = 0$
   (E) $x - y - 12 = 0$

14. In the figure to the right, the circle has a radius of 7.07. AC is a diameter, and $AB = BC$. The area of triangle ABC is

   (A) 100
   (B) 200
   (C) 50
   (D) 75
   (E) 25

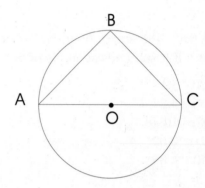

15. In the figure to the right, we have a cube with sides with a length of 4 each.

   The area of rectangle ABGH is

   (A) 15.28
   (B) 16.52
   (C) 18.9
   (D) 22.62
   (E) 19.05

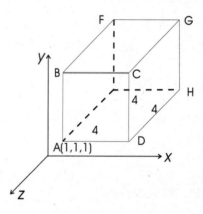

16. If $x - 2y = 0$ and $y^2 - 4x = 0$ (for all $x > 0$), $x =$

   (A) 13.65
   (B) 15
   (C) 16
   (D) 17.28
   (E) 18.5

17. For all values of $\theta \leq \frac{\pi}{2}$, $(\sec 2\theta)(\cos 2\theta) =$

    (A) 0.5
    (B) 1
    (C) 0.65
    (D) 0.38
    (E) 0

18. If the distance between a 13-foot ladder and a vertical wall is 5 feet along the ground, how high can a person climb if the ladder is inclined against the wall?

    (A) 18 feet
    (B) 65 feet
    (C) $\frac{13}{5}$ feet
    (D) 8 feet
    (E) 12 feet

19. What is the probability that the first 2 draws from a pack of cards are clubs and the third is a spade?

    (A) $\frac{2,028}{132,600}$
    (B) $\frac{38}{132,600}$
    (C) $\frac{1,716}{132,600}$
    (D) $\frac{2,028}{140,608}$
    (E) $\frac{1,716}{140,608}$

DIAGNOSTIC TEST 2

MATHEMATICS LEVEL IIC TEST—Continued

USE THIS SPACE FOR SCRATCH WORK

20. In triangle ABC, AB = 1.5(AC) and BC = 10. The perimeter of the triangle is 25. If ∠ACB = 35°, what is the value of ∠ABC?

   (A) 59.4°
   (B) 22.5°
   (C) 35°
   (D) 42.6°
   (E) 24°

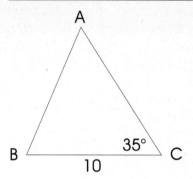

21. The graph of $f(x) = 10 - 4e^{-2x}$ is shown to the right. What is the area of triangle ABC if OA = AB?

   (A) 25
   (B) 60
   (C) 45
   (D) 50
   (E) 30

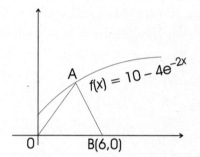

22. A purse contains a total of 15 coins, all of which are either dimes or nickels. If the value of the collection is $1.20, how many are dimes and how many are nickels?

   (A) 10 dimes and 5 nickels
   (B) 7 dimes and 8 nickels
   (C) 11 dimes and 4 nickels
   (D) 9 dimes and 6 nickels
   (E) 8 dimes and 7 nickels

23. The value of k so that the sequence $k - 1$, $k + 3$, $3k - 1$ forms an arithmetic progression is

   (A) 6
   (B) 5
   (C) 4
   (D) 3
   (E) 2

24. In triangle *OAB*, if *OA* = *OB*, *AB* is parallel to *CD* and the height of the triangle *OCD* is 3, the area of the triangle *OAB* is

(A) 24
(B) 48
(C) 18
(D) 32
(E) 28

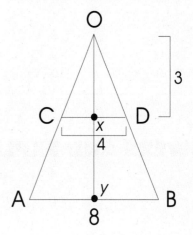

25. If $\sec\theta = 1.414$, $\sin\theta =$

(A) 0.5
(B) 0.71
(C) 0.86
(D) 1.414
(E) 0.67

DIAGNOSTIC TEST 2

Quick Score Answers

1. B	6. D	11. B	16. C	21. E
2. D	7. A	12. B	17. B	22. D
3. C	8. C	13. A	18. E	23. C
4. C	9. D	14. C	19. A	24. A
5. E	10. E	15. D	20. B	25. B

ANSWERS AND EXPLANATIONS

1. **The correct answer is (B).** We first take the common denominator in this problem by multiplying and dividing the entire expression by (bc) and then multiply out the b. Thus we have

$$b\left(\frac{3a}{b} + \frac{2}{c}\right) = b\left[\frac{3a}{b}\left(\frac{bc}{bc}\right) + \frac{2}{c}\left(\frac{bc}{bc}\right)\right] = b\left[\frac{3ac}{bc} + \frac{2b}{bc}\right] = b\left[\frac{3ac + 2b}{bc}\right]$$

$$= \frac{3ac + 2b}{c}.$$

2. **The correct answer is (D).** In this problem, we first use the property of exponents, $u^p \cdot u^q = u^{p+q}$, and once we have a more simplified expression, we take natural logs on both sides and use the property $\ln x^k = k \ln x$. Thus, we have

$$2^x 2^y = 3^z$$
$$\Rightarrow 2^{x+y} = 3^z$$
$$\Rightarrow \ln(2)^{x+y} = \ln(3)^z$$
$$\Rightarrow (x+y)\ln 2 = z\ln 3$$
$$\Rightarrow \frac{x+y}{z} = \frac{\ln 3}{\ln 2} = 1.585.$$

3. **The correct answer is (C).** For this problem, we recall that $i^2 = -1$. Multiplying out the two expressions, we have

$(a + 2i)(b - i)$
$= ab + 2bi - ai - 2i^2$
$= ab + (2b - a)i - 2(-1)$
$= ab + (2b - a)i + 2.$

4. **The correct answer is (C).** For this problem, we use the property of quadrilaterals, which states that the sum of internal angles of a quadrilateral equals 360°. Thus, $A + B + C + D = 360$. But we have been given that $A = B$, $C = \left(\frac{1}{2}\right)A$, and $D = A + 10$. This gives us

$$A + A + \left(\frac{1}{2}\right)A + (A + 10) = 360$$
$$\Rightarrow A = 100°$$
$$\Rightarrow B = 100°, C = 50°, \text{ and } D = 110°.$$

Peterson's SAT II Success:
Mathematics IC and IIC

ANSWERS AND EXPLANATIONS

5. **The correct answer is (E).** The profit made on selling the item is $(165 - 150) = \$15$. The percentage of profit in terms of the original price is given as $\left(\dfrac{15}{150}\right)100 = 10$ percent.

6. **The correct answer is (D).** We have $f(x) = \dfrac{x^2 - 1}{x + 3}$ and $g(f(5)) = 21$. To find $g(t)$, we first find the value of $f(5)$. This is obtained by plugging in 5 for x in the expression for $f(x)$. Thus, $f(5) = \dfrac{(5)^2 - 1}{5 + 3} = 3$.

   This gives us $g(t = 3) = 21$. Now, with this information, we plug in 3 for t in the 5 choices given to us and see which of them gives us 21 for an answer. We observe that for choice (D), we have $t^2 + 3t + 3 = (3)^2 + 3(3) + 3 = 21$. Thus, $g(t) = t^2 + 3t + 3$.

7. **The correct answer is (A).** The volume of a pyramid is given as $V = \dfrac{1}{3}Ah$, where A = base area and h = height of the pyramid. The volume of a pyramid with a square base with a side of 6 and a height of 12 is $V_1 = \dfrac{1}{3}(36)(12) = 144$. A similar pyramid with a height of 3 will have a base side equal to $3 \cdot \dfrac{6}{12} = 1.5$. The volume of a pyramid with a square base with a side of 1.5 and a height of 3 is $V_2 = \dfrac{1}{3}(2.25)(3) = 2.25$. Therefore, the volume of the solid obtained by cutting off a pyramid with a height of 3 from a bigger pyramid with a height of 12 is $V = V_1 - V_2 = 144 - 2.25 = 141.75$.

8. **The correct answer is (C).** We have been given a function $f(x) = ax^2 + bx + c$. To find a and b, we use the information given to us. We have

   $f(0) = 1 \Rightarrow a(0)^2 + b(0) + c = 1 \Rightarrow c = 1$, and

   $f(-1) = 3 \Rightarrow a(-1)^2 + b(-1) + c = 3 \Rightarrow a - b + 1 = 3 \Rightarrow a - b = 2$.

9. **The correct answer is (D).** To find the domain of the function $f(x) = \dfrac{x - 4}{\sqrt{1 - 2x}}$, we observe that the function is defined only if the value under the radical sign in the denominator is greater than zero. This is because we cannot have a division by zero, which rules out a zero in the denominator, and we cannot have a negative value inside the radical sign, since that would make it a complex number. Therefore, the function is defined only if $(1 - 2x) > 0$, giving us $x < \dfrac{1}{2}$, as the possible value.

10. **The correct answer is (E).** Since the bug traverses a circular path in half an hour at the rate of 62.84 inches/hr, it covers a complete circle of circumference $\left(\frac{62.84}{2}\right) = 31.42$ inches in the half hour. The radius of the circle, having a circumference of 31.42, can be obtained using the formula $C = 2\pi r$, where C = circumference and r = radius of the circle. Plugging in the value of C and solving for r, we have $31.42 = 2\pi r \Rightarrow r = 5$ inches.

11. **The correct answer is (B).** The length of an arc subtending an angle θ at the center of a circle of radius r is $L = \frac{\theta}{360}(2\pi r)$. Using this formula, the length of an arc subtending 60° at the center of a circle with a radius of 5 is $L = \frac{60}{360}(2\pi)(5) = 5.24$.

12. **The correct answer is (B).** For the function
$$f(x) = x^2 - 6x, \quad x \leq 2$$
$$= 2x - 1, \quad x > 2,$$
the value of $f(3)$ is defined by $f(x) = 2x - 1$. Thus, plugging in 3 for x, we have $f(3) = 2(3) - 1 = 5$.

13. **The correct answer is (A).** We use the property of tangents to circles here, which states that the tangent to a circle at any point on the circle is at right angles to the line joining that point and the center of the circle.

    The slope of the line joining the center at point (3, 3) and the point (6, 6) at which we have the tangent line touching the circle is $\frac{6-3}{6-3} = 1$. Now, since the slopes, $m1$ and $m2$, of two lines perpendicular to each other are related such that $(m1)(m2) = -1$, we get the slope of the tangent line as -1. Thus, the equation of the tangent line that has slope -1 and passing through point (6, 6) is $(y - 6) = (-1)(x - 6) \Rightarrow x + y - 12 = 0$.

14. **The correct answer is (C).** For a circle, we know that any point on the circumference forms a right-angled triangle with a diameter of the circle such that the diameter is the hypotenuse of the right-angled triangle. Therefore, for the given problem, the hypotenuse (which is the diameter of the circle) is of length 14.14. Also, it is given that $AB = BC$. Using this information and the Pythagorean theorem, we have $(AC)^2 = (AB)^2 + (BC)^2 \Rightarrow AB = BC = 10$.

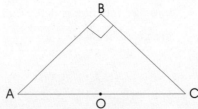

Therefore, the area of the triangle is $\left(\frac{1}{2}\right)(AB)(BC) = 50$ square units.

ANSWERS AND EXPLANATIONS

15. **The correct answer is (D).** For the given cube with sides of 4 each, the coordinates of points B, G, and H are (1, 5, 1), (5, 5, −3), and (5, 1, −3), respectively. To obtain the area of rectangle ABGH, we need to calculate the length of sides AB (or GH) and BG (or AH). The distance between two points $(x1, y1, z1)$ and $(x2, y2, z2)$ is given as $D = \sqrt{(x2-x1)^2 + (y1-y2)^2 + (z1-z2)^2}$.

    The distance between points A and B is the same as the length of a side of the cube, which is 4 units. To obtain the distance between points B(1, 5, 1) and G(5, 5, −3), we use the above distance formula. Thus,
    $BG = \sqrt{(5-1)^2 + (5-5)^2 (-3-1)^2} = \sqrt{32} = 5.66$.

    Therefore, the area of rectangle ABGH is $(AB)(BG) = (4)(5.66) = 22.62$.

16. **The correct answer is (C).** To obtain the value of x, we solve the two equations $x - 2y = 0$ and $y^2 - 4x = 0$ simultaneously. Plugging in $x = 2y$ in the second equation, we have $y^2 - 4(2y) = 0 \Rightarrow y = 8$. This gives us $x = 16$.

17. **The correct answer is (B).** By definition, $\cos 2\theta = \dfrac{1}{\sec 2\theta} \Rightarrow (\sec 2\theta)(\cos 2\theta) = (\sec 2\theta)\left(\dfrac{1}{\sec 2\theta}\right) = 1$.

18. **The correct answer is (E).** We refer to the figure below for this problem. The height that a person can climb using the ladder can be obtained using the Pythagorean theorem as $h = \sqrt{(13)^2 - (5)^2} = 12$.

    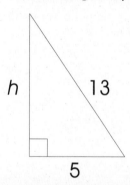

19. **The correct answer is (A).** A pack of cards contains 13 each of clubs, spades, hearts, and diamonds. The probability of obtaining a club on the first draw is $\dfrac{13}{52}$, since there are 13 clubs to choose from the pack of 52 cards. Now, without replacing the first card, if we have to draw another club, we have 12 clubs to pick from the pack that has 51 cards. This has a probability of $\dfrac{12}{51}$. Finally, without replacing the first two cards, if we need to draw a spade from the remaining cards in the pack, we have a probability of $\dfrac{13}{50}$. Therefore, the total probability of drawing two clubs and a spade successively without replacing the cards will be $\left(\dfrac{13}{52}\right)\left(\dfrac{12}{51}\right)\left(\dfrac{13}{50}\right) = \dfrac{2{,}028}{132{,}600}$.

20. **The correct answer is (B).** Since the perimeter of the triangle is given to be equal to 25, we have

$$l(AB) + l(BC) + l(AC) = 25$$
$$\Rightarrow \frac{3}{2}l(AC) + 10 + l(AC) = 25$$
$$\Rightarrow l(AC) = 6$$
$$\Rightarrow (AB) = 9.$$

Now, we make use of the Sine Rule of Triangles, which is as follows:

$$\frac{\sin A}{l(BC)} = \frac{\sin B}{l(AC)} = \frac{\sin C}{l(AB)}.$$

Thus,

$$\frac{\sin 35}{9} = \frac{\sin B}{6} \Rightarrow \sin B = \frac{6(\sin 35)}{9} = 0.38 \Rightarrow B = \sin^{-1}(0.38) = 22.5°.$$

Therefore, $\angle ABC = 22.5°$.

21. **The correct answer is (E).** Since triangle OAB is an isosceles triangle, we know that the perpendicular bisector of side OB will pass through the vertex A. If AX is the perpendicular bisector of side OB, then the coordinates of point X will be $(3, 0)$. To calculate the area of the triangle, we need to find the height of the triangle, which happens to be the y-coordinate of point A. The x-coordinate of point A is the same as that of point X, which is 3. To get the y-coordinate of point A, we notice that the point lies on the curve $y = 10 - 4e^{-2x}$. Plugging in 3 for x, we have $y = 10 - 4e^{-2(3)} \approx 10$.

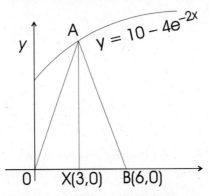

Thus, the height of the triangle $= AX = 10$.

The area of triangle $OAB = \left(\frac{1}{2}\right)(AX)(OB) = \left(\frac{1}{2}\right)(10)(6) = 30$.

22. **The correct answer is (D).** Let there be x dimes and y nickels. Then we have $x + y = 15$ (the total number of coins is 15), $(0.1)x + (0.05)y = 1.20$ (the total value of the collection is $1.20). Solving the above two equations simultaneously, we have $x = 9$ and $y = 6$.

23. **The correct answer is (C).** If the sequence $(k - 1)$, $(k + 3)$, $(3k - 1)$ is an arithmetic progression, then, by definition of an arithmetic progression, we have $(k + 3) - (k - 1) = (3k - 1) - (k + 3)$. Solving the above equation for k, we get $k = 4$.

24. **The correct answer is (A).** In the figure below, since AB is parallel to CD, the two triangles OXB and OYD are similar. Also, since the triangles are isosceles triangles, OX and OY are the perpendicular bisectors of sides AB and CD, respectively. By property of similar triangles, we have
$\frac{OD}{OB} = \frac{OX}{OY} = \frac{XD}{YB}$. This gives us $\frac{2}{4} = \frac{3}{OY} \Rightarrow OY = 6$, since $XD = 2$ (which is half of CD) and $YB = 4$ (which is half of AB).

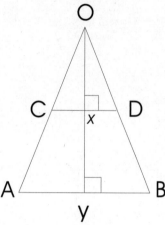

Therefore, the area of triangle OAB is Area $= \left(\frac{1}{2}\right)(OY)(AB) = \left(\frac{1}{2}\right)(6)(8) = 24$.

25. **The correct answer is (B).** If $\sec\theta = 1.414$, $\cos\theta = \frac{1}{\sec\theta} = \frac{1}{1.414}$. Using the trigonometric identity, $\sin^2\theta + \cos^2\theta = 1$, we have $\sin\theta = \sqrt{1 - \cos^2\theta} = \sqrt{1 - \left(\frac{1}{1.414}\right)^2} = 0.71$.

Mathematics IC and IIC Review

MATHEMATICS LEVEL IC AND IIC REVIEW

OVERVIEW

The following chapter contains a review of all of the mathematics tested on both the Level IC and the Level IIC Subject Tests. While these two tests share many topics, the Level IIC test, in general, emphasizes more advanced mathematics. Therefore, for example, while a student taking Level IC must know some basic right triangle trigonometry, a student taking Level IIC must know more detailed information on trigonometric functions. In addition, there are several math topics, such as exponential functions, logarithms, and complex numbers, that appear only on the Level IIC test.

Students preparing for Level IIC should study *all* of the math review topics in this chapter. Those taking Level IC should start at the beginning. As they proceed through the review, they will notice that, from time to time, there will be topics clearly labeled as "for Level IIC only." Such topics will not be covered on the Level IC test, so they can skip over them.

ARITHMETIC

While neither the Level IC nor the Level IIC test contains any problems that specifically test arithmetic skills per se, a solid knowledge of basic arithmetic will certainly be needed throughout both tests. Therefore, it is a good idea to begin your review by making sure that you are familiar with all of the topics in this section.

PROPERTIES OF NUMBERS

Systems of Numbers

Typically, the numbers that are used on the SAT II tests are *real numbers*. The only exception to this is on the Level IIC test, which may contain a small number of problems involving imaginary and complex numbers. These number systems are discussed in a later section. Thus, for the time being, we will restrict our focus to the real number system.

In order to understand the real number system, it is easiest to begin by looking at some familiar systems of numbers that lie within the real number system.

The numbers that are used for counting

1, 2, 3, 4, 5, ...

are called the *natural numbers,* the *counting numbers,* or, most commonly, the *positive integers*. The positive integers, together with the number 0, are called the set of *whole numbers*. Then, the positive integers, together with 0 and the *negative integers*

−1, −2, −3, −4, −5, ...

make up the set of *integers*. Thus, the set of integers contains the numbers

... −5, −4, −3, −2, −1, 0, 1, 2, 3, 4, 5, ...

A real number is said to be a *rational number* if it can be written as the ratio of two integers, where the denominator is not 0. Thus, for example, numbers such as

$$-16, \frac{3}{4}, \frac{-5}{6}, 0, 49, 13\frac{7}{12}$$

are rational numbers. Clearly, then, all integers and fractions are rational numbers. Percents and decimal numbers are rational as well, since they can also be written as the ratio of two integers. For example, $75\% = \frac{3}{4}$, and $9.375 = \frac{3}{8}$.

Any real number that cannot be expressed as the ratio of two integers is called an *irrational number*. The most common irrational numbers that you will see on your test are square roots, such as $\sqrt{7}$ or $-\sqrt{13}$, and the number π, which represents the ratio of the circumference of a circle to its diameter.

Finally, the set of rational numbers, together with the set of irrational numbers, is called the set of *real numbers*.

Example

The number -293 is an integer. It is also rational since it can be written as $\frac{-293}{1}$ and is, of course, real.

The number $\frac{5}{8}$ is rational and real, and the number 15.0625 is also rational and real since it can be written as $15\frac{1}{16}$ or $\frac{241}{16}$.

The number $\sqrt{5}$ is irrational and real.

ROUNDING OF NUMBERS

From time to time, a test question will ask you to round an answer to a specific decimal place. The rules for the rounding of numbers are very simple. In the case of whole numbers, begin by locating the digit to which the number is being rounded. Then, if the digit just to the right is 0, 1, 2, 3, or 4, leave the located digit alone. Otherwise, increase the located digit by 1. In either case, replace all digits to the right of the one located with 0s.

When rounding decimal numbers, the rules are similar. Again, begin by locating the digit to which the number is being rounded. As before, if the digit just to the right is 0, 1, 2, 3, or 4, leave the located digit alone. Otherwise, increase the located digit by 1. Finally, drop all the digits to the right of the one located.

Examples

Round the following numbers as indicated:

1. 7,542 to the nearest tenth

    Begin by locating the tens' digit, which is a 4. The number to the right of the 4 is a 2. Thus, drop the 2 and replace it with a 0, yielding 7,540.

2. 495,597 to the nearest hundred

    Begin by locating the hundreds' digit, which is a 5. The number to the right of the 5 is a 9. Thus, increase the hundred's digit by 1, making it a 6. Replace the tens' and units' digits with 0s, yielding 495,600.

3. 893.472 to the nearest tenth

    The tenths' digit is 4. The digit just to the right of it is 7, so increase the tenth's digit by 1, making it a 5. Drop the two digits to the right of this. The answer is 893.5

4. .0679 to the nearest thousandth

    Following the rules above, we obtain .068.

PROPERTIES OF NUMBERS PROBLEMS

QUIZ

1. Classify each of the following numbers as whole, integer, rational, irrational, and real.

   (A) -7

   (B) $\frac{1}{7}$

   (C) $5\frac{2}{3}$

   (D) 0

   (E) $\sqrt{13}$

2. Round each of the numbers below to the indicated number of decimal places.

   (A) 57,380 to the nearest hundred
   (B) 1,574,584 to the nearest hundred thousand
   (C) 847.235 to the nearest hundredth
   (D) 9.00872 to the nearest thousandth

SOLUTIONS

1. (A) -7 is real, rational, and an integer.

   (B) $\frac{1}{7}$ is real and rational.

   (C) $5\frac{2}{3}$ can be written as $\frac{17}{3}$ and is, thus, real and rational.

   (D) 0 is real, rational, an integer, and a whole number.

   (E) $\sqrt{13}$ is real and irrational.

2. (A) Begin by locating the hundreds' digit, which is 3. The digit to the right of it is 8, so increase the hundreds' digit by 1, and replace all digits to the right with 0s. The answer is 57,400.

   (B) The hundred thousands digit is 5. The digit to the right' of it is 7, so increase the 5 by 1 and replace all digits to the right with 0s. The answer is 1,600,000.

   (C) The hundredth's digit is 3. The digit just to the right of it is 5, so increase the hundredth's digit by 1, making it a 4. Drop the digit to the right of this. The answer is 847.24.

   (D) The thousandth's digit is 8. The digit just to the right of it is 7, so increase the thousandths' digit by 1, making it a 9. Drop the digits to the right of this. The answer is 9.009.

Peterson's: www.petersons.com

Whole Numbers

As we have already seen, the set of positive integers (natural numbers, counting numbers) can be written as the set {1, 2, 3, 4, 5, ...}. The set of positive integers, together with the number 0, are called the set of *whole numbers*, and can be written as {0, 1, 2, 3, 4, ...}. (The notation { } means "set" or collection, and the three dots after the number 5 indicate that the list continues without end.)

Place Value

Whole numbers are expressed in a system of tens, called the *decimal* system. Ten *digits*—0, 1, 2, 3, 4, 5, 6, 7, 8, and 9—are used. Each digit differs not only in *face* value but also in *place* value, depending on where it stands in the number.

Example 1

237 means:

$(2 \cdot 100) + (3 \cdot 10) + (7 \cdot 1)$

The digit 2 has face value 2 but place value of 200.

Example 2

35,412 can be written as:

$(3 \cdot 10,000) + (5 \cdot 1,000) + (4 \cdot 100) + (1 \cdot 10) + (2 \cdot 1)$

The digit in the last place on the right is said to be in the units or ones place, the digit to the left of that in the ten's place, the next digit to the left of that in the hundred's place, and so on.

When we take a whole number and write it out as in the two examples above, it is said to be written in *expanded form*.

Odd and Even Numbers

A whole number is *even* if it is divisible by 2; it is *odd* if it is not divisible by 2. Zero is thus an even number.

Example

2, 4, 6, 8, and 320 are even numbers; 3, 7, 9, 21, and 45 are odd numbers.

Prime Numbers

The positive integer p is said to be a prime number (or simply *a prime*) if $p \neq 1$ and the only positive divisors of p are itself and 1. The positive integer 1 is called a *unit*. The first ten primes are 2, 3, 5, 7, 11, 13, 17, 19, 23, and 29. All other positive integers that are neither 1 nor prime are *composite numbers*. Composite numbers can be *factored*, that is, expressed as products of their divisors or factors; for example, $56 = 7 \cdot 8 = 7 \cdot 4 \cdot 2$. In particular, composite numbers can be expressed as products of their *prime* factors in just one way (except for order).

To factor a composite number into its prime factors, proceed as follows. First, try to divide the number by the prime number 2. If this is successful, continue to divide by 2 until an odd number is obtained. Then, attempt to divide the last quotient by the prime number 3 and by 3 again, as many times as possible. Then move on to dividing by the prime number 5 and other successive primes until a prime quotient is obtained. Express the original number as a product of all its prime divisors.

Example

Find the prime factors of 210.

$2 \overline{)210}$
$3 \overline{)105}$
$5 \overline{)\ 35}$
$\qquad 7$

Therefore:

$210 = 2 \cdot 3 \cdot 5 \cdot 7$ (written in any order)
and 210 is an integer multiple of 2, of 3, of 5, and of 7.

Consecutive Whole Numbers

Numbers are consecutive if each number is the successor of the number that precedes it. In a consecutive series of whole numbers, an odd number is always followed by an even number, and an even number by an odd. If three consecutive whole numbers are given, either two of them are odd and one is even or two are even and one is odd.

Example 1

7, 8, 9, 10, and 11 are consecutive whole numbers.

Example 2

8, 10, 12, and 14 are consecutive even numbers.

Example 3

21, 23, 25, and 27 are consecutive odd numbers.

Example 4

21, 23, and 27 are *not* consecutive odd numbers because 25 is missing.

THE NUMBER LINE

A useful method of representing numbers geometrically makes it easier to understand numbers. It is called the *number line*. Draw a horizontal line, considered to extend without end in both directions. Select some point on the line and label it with the number 0. This point is called the *origin*. Choose some convenient distance as a unit of length. Take the point on the number line that lies one unit to the right of the origin and label it with the number 1. The point on the number line that is one unit to the right of 1 is labeled 2, and so on. In this way, every whole number is associated with one point on the line, but it is not true that every point on the line represents a whole number.

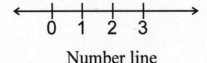

Number line

PROPERTIES OF NUMBERS

Ordering of Whole Numbers

On the number line, the point representing 8 lies to the right of the point representing 5, and we say $8 > 5$ (read "8 is greater than 5"). One can also say $5 < 8$ ("5 is less than 8"). For any two whole numbers a and b, there are always three possibilities:

$$a < b, \quad a = b, \quad \text{or} \quad a > b.$$

If $a = b$, the points representing the numbers a and b coincide on the number line.

OPERATIONS WITH WHOLE NUMBERS

The basic operations on whole numbers are addition (+), subtraction (−), multiplication (· or ×), and division (÷). These are all *binary* operations—that is, one works with two numbers at a time in order to get a unique answer. The operations of addition and multiplication on whole numbers are said to be *closed* because the answer in each case is also a whole number. The operations of subtraction and division on whole numbers are not closed because the unique answer is not necessarily a member of the set of whole numbers.

Examples

$$3 + 4 = 7 \quad \text{a whole number}$$
$$4 \cdot 3 = 12 \quad \text{a whole number}$$
$$2 - 5 = -3 \quad \text{not a whole number}$$
$$3 \div 8 = \frac{3}{8} \quad \text{not a whole number}$$

Addition

If addition is a binary operation, how are three numbers—say, 3, 4, and 8—added? One way is to write:

$$(3 + 4) + 8 = 7 + 8 = 15$$

Another way is to write:

$$3 + (4 + 8) = 3 + 12 = 15$$

The parentheses merely group the numbers together. The fact that the same answer, 15, is obtained either way illustrates the *associative property* of addition:

$$(r + s) + t = r + (s + t)$$

MATHEMATICS IC AND IIC REVIEW

The order in which whole numbers are added is immaterial—that is, 3 + 4 = 4 + 3. This principle is called the *commutative property* of addition. Most people use this property without realizing it when they add a column of numbers from the top down and then check their results by beginning over again from the bottom. (Even though there may be a long column of numbers, only two numbers are added at a time.)

If 0 is added to any whole number, the whole number is unchanged. Zero is called the *identity element* for addition.

Subtraction

Subtraction is the inverse of addition. The order in which the numbers are written is important; there is no commutative property for subtraction.

$$4 - 3 \neq 3 - 4$$

The $\neq$ is read "does not equal."

Multiplication

Multiplication is a commutative operation:

$$43 \cdot 73 = 73 \cdot 43$$

The result or answer in a multiplication problem is called the *product*.

If a number is multiplied by 1, the number is unchanged; the *identity element* for multiplication is 1.

Zero times any number is 0:

$$42 \cdot 0 = 0$$

Multiplication can be expressed with several different symbols:

$$9 \cdot 7 \cdot 3 = 9 \times 7 \times 3 = 9(7)(3)$$

Besides being commutative, multiplication is *associative*:

$$(9 \cdot 7) \cdot 3 = 63 \cdot 3 = 189$$

and

$$9 \cdot (7 \cdot 3) = 9 \cdot 21 = 189$$

A number can be quickly multiplied by 10 by adding a zero to the right of the number. Similarly, a number can be multiplied by 100 by adding two zeros to the right:

$$38 \cdot 10 = 380$$

and

$$100 \cdot 76 = 7{,}600$$

Division

Division is the inverse of multiplication. It is not commutative:

$$8 \div 4 \neq 4 \div 8$$

The parts of a division example are named as follows:

$$\text{divisor}\overline{)\text{dividend}}^{\text{quotient}}$$

If a number is divided by 1, the quotient is the original number.

Division by 0 is not defined (has no meaning). Zero divided by any number other than 0 is 0:

$$0 \div 56 = 0$$

Divisors and Multiples

The whole number b *divides* the whole number a if there exists a whole number k such that $a = bk$. The whole number a is then said to be an integer *multiple* of b, and b is called a *divisor* (or *factor*) of a.

Example 1

3 divides 15 because $15 = 3 \cdot 5$. Thus, 3 is a divisor of 15 (and so is 5), and 15 is an integer multiple of 3 (and of 5).

Example 2

3 does not divide 8 because $8 \neq 3k$ for a whole number k.

Example 3

Divisors of 28 are 1, 2, 4, 7, 14, and 28.

Example 4

Multiples of 3 are 3, 6, 9, 12, 15, . . .

QUIZ

WHOLE NUMBERS PROBLEMS

1. What are the first seven positive multiples of 9?

2. What are the divisors of 60?

3. Find all of the factors of 47.

4. Express 176 as a product of prime numbers.

5. Find all of the common prime factors of 30 and 105.

6. Give an example to show that subtraction on the set of real numbers is not commutative.

7. List all of the prime numbers between 50 and 90.

8. Write the number 786,534 in expanded notation.

9. Which property is illustrated by the following statement?

    $(9 \times 7) \times 5 = 9 \times (7 \times 5)$

10. Which property is illustrated by the following statement?

    $(16 + 18) + 20 = (18 + 16) + 20$

11. Which property is illustrated by the following statement?

    $(3 + 7) + 8 = 3 + (7 + 8)$

12. In each of the statements below, replace the # with <, >, or = to make a true statement.

    (A) -12 # 13

    (B) $\dfrac{1}{16}$ # 0.0625

    (C) $3\dfrac{1}{2}$ # $3\dfrac{2}{5}$

SOLUTIONS

1. 9, 18, 27, 36, 45, 54, 63

2. The divisors of 60 are 1, 2, 3, 4, 5, 6, 10, 12, 15, 20, 30, and 60.

3. 1 and 47 are the only factors.

4. $176 = 2 \times 88 = 2 \times 2 \times 44 = 2 \times 2 \times 2 \times 22 = 2 \times 2 \times 2 \times 2 \times 11$

5. 30 can be factored as $2 \times 3 \times 5$. 105 can be factored as $3 \times 5 \times 7$. Thus, the common prime factors are 3 and 5.

6. $4 - 5 \neq 5 - 4$

7. The prime numbers between 50 and 90 are 53, 59, 61, 67, 71, 73, 79, 83, and 89.

8. $786{,}534 = 7(100{,}000) + 8(10{,}000) + 6(1{,}000) + 5(100) + 3(10) + 4$

9. The associative property of multiplication

10. The commutative property of addition

11. The associative property of addition

12. (A) $-12 < 13$
    (B) $\frac{1}{16} = 0.0625$
    (C) $3\frac{1}{2} > 3\frac{2}{5}$

FRACTIONS

Definitions

If a and b are whole numbers and $b \neq 0$, the symbol $\frac{a}{b}$ (or a/b) is called a fraction. The upper part, a, is called the *numerator,* and the lower part, b, is called the *denominator*. The denominator indicates into how many parts something is divided, and the numerator tells how many of these parts are taken. A fraction indicates division:

$$\frac{7}{8} = 8\overline{)7}$$

If the numerator of a fraction is 0, the value of the fraction is 0. If the denominator of a fraction is 0, the fraction is not defined (has no meaning):

$$\frac{0}{17} = 0$$

$\frac{17}{0}$ is not defined (has no meaning)

If the denominator of a fraction is 1, the value of the fraction is the same as the numerator:

$$\frac{18}{1} = 18$$

If the numerator and denominator are the same number, the value of the fraction is 1:

$$\frac{7}{7} = 1$$

Equivalent Fractions

Fractions that represent the same number are said to be *equivalent*. If m is a counting number and $\frac{a}{b}$ is a fraction, then: $\frac{m \times a}{m \times b} = \frac{a}{b}$ because $\frac{m}{m} = 1$ and $1 \times \frac{a}{b} = \frac{a}{b}$

Example

$$\frac{2}{3} = \frac{4}{6} = \frac{6}{9} = \frac{8}{12}$$

These fractions are all equivalent.

Inequality of Fractions

If two fractions are not equivalent, one is smaller than the other. The ideas of "less than" and "greater than" were previously defined and used for whole numbers.

For the fractions $\frac{a}{b}$ and $\frac{c}{b}$:

$\frac{a}{b} < \frac{c}{b}$ if $a < c$ and if $b > 0$

That is, if two fractions have the same denominator, the one with the smaller numerator has the smaller value.

If two fractions have different denominators, find a common denominator by multiplying one denominator by the other. Then use the common denominator to compare numerators.

Example 1

Which is smaller, $\frac{5}{8}$ or $\frac{4}{7}$?

$8 \cdot 7 = 56$ = common denominator

$\frac{5}{8} \times \frac{7}{7} = \frac{35}{56}$ $\quad$ $\frac{4}{7} \times \frac{8}{8} = \frac{32}{56}$

Since $32 < 35$,

$\frac{32}{56} < \frac{35}{56}$ and $\frac{4}{7} < \frac{5}{8}$

Example 2

Which of the fractions, $\frac{2}{5}, \frac{3}{7}$, or $\frac{4}{11}$, is the largest?

We begin by comparing the first two fractions. Since $\frac{2}{5} = \frac{14}{35}$ and $\frac{3}{7} = \frac{15}{35}$, we can see that $\frac{3}{7}$ is bigger. Now, we compare $\frac{3}{7}$ to $\frac{4}{11}$. Since $\frac{3}{7} = \frac{33}{77}$ and $\frac{4}{11} = \frac{28}{77}$, we can see that $\frac{3}{7}$ is the biggest of the three fractions.

Reducing to Lowest Terms

The principle that

$$\frac{m \times a}{m \times b} = \frac{a}{b}$$

can be particularly useful in reducing fractions to lowest terms. Fractions are expressed in *lowest terms* when the numerator and denominator have no common factor except 1. To reduce a fraction to an equivalent fraction in lowest terms, express the numerator and denominator as products of their prime factors. Each time a prime appears in the numerator over the same prime in the denominator, $\frac{p}{p}$, substitute its equal value, 1.

Example 1

Reduce $\frac{30}{42}$ to an equivalent fraction in lowest terms:

$$\frac{30}{42} = \frac{2 \cdot 3 \cdot 5}{2 \cdot 3 \cdot 7} = 1 \cdot 1 \cdot \frac{5}{7} = \frac{5}{7}$$

In practice, this can be done even more quickly by dividing the numerator and the denominator by any number, prime or not, that will divide both evenly. Repeat this process until there is no prime factor remaining that is common to both the numerator and the denominator:

$$\frac{30}{42} = \frac{15}{21} = \frac{5}{7}$$

Example 2

Reduce $\frac{77}{197}$ to an equivalent fraction in lowest terms:

$$\frac{77}{197} = \frac{7 \times 11}{3 \times 5 \times 13}$$

Since the numerator and the denominator have no common factors, the fraction is already in lowest terms.

PROPERTIES OF NUMBERS

QUIZ

FRACTIONS PROBLEMS

In the following problems, perform the indicated operations and reduce your answers to lowest terms.

1. $\dfrac{2}{15} + \dfrac{2}{3} =$

2. $\dfrac{4}{5} - \dfrac{2}{13} =$

3. $\dfrac{3}{8} \times \dfrac{4}{21} =$

4. $\dfrac{2}{3} \times \dfrac{12}{8} =$

5. $\dfrac{2}{3} \div \dfrac{5}{6} =$

6. $\dfrac{3}{4} \div \dfrac{7}{8} =$

7. $2\dfrac{3}{5} + 7\dfrac{3}{5} =$

8. $9\dfrac{1}{5} - 3\dfrac{1}{4} =$

9. $\dfrac{6}{7} \times \dfrac{3}{4} \times \dfrac{2}{3} =$

10. $6 \times \dfrac{2}{3} \times 2\dfrac{5}{6} =$

11. $2\dfrac{2}{3} \div 1\dfrac{7}{9} =$

12. $6\dfrac{2}{3} \times 1\dfrac{4}{5} =$

SOLUTIONS

1. $\dfrac{2}{15} + \dfrac{2}{3} = \dfrac{2}{15} + \dfrac{10}{15} = \dfrac{12}{15} = \dfrac{4}{5}$

2. $\dfrac{4}{5} - \dfrac{2}{13} = \dfrac{52}{65} - \dfrac{10}{65} = \dfrac{42}{65}$

3. $\dfrac{3}{8} \times \dfrac{4}{21} = \dfrac{4}{8} \times \dfrac{3}{21} = \dfrac{1}{2} \times \dfrac{1}{7} = \dfrac{1}{14}$

4. $\dfrac{2}{3} \times \dfrac{12}{8} = \dfrac{12}{3} \times \dfrac{2}{8} = \dfrac{4}{1} \times \dfrac{1}{4} = 1$

5. $\dfrac{2}{3} \div \dfrac{5}{6} = \dfrac{2}{3} \times \dfrac{6}{5} = \dfrac{6}{3} \times \dfrac{2}{5} = \dfrac{2}{1} \times \dfrac{2}{5} = \dfrac{4}{5}$

6. $\dfrac{3}{4} \div \dfrac{7}{8} = \dfrac{3}{4} \times \dfrac{8}{7} = \dfrac{8}{4} \times \dfrac{3}{7} = \dfrac{2}{1} \times \dfrac{3}{7} = \dfrac{6}{7}$

7. $2\dfrac{3}{5} + 7\dfrac{3}{5} = \dfrac{13}{5} + \dfrac{38}{5} = \dfrac{51}{5} = 10\dfrac{1}{5}$

8. $9\dfrac{1}{5} - 3\dfrac{1}{4} = 8\dfrac{6}{5} - 3\dfrac{1}{4} = 8\dfrac{24}{20} - 3\dfrac{5}{20} = 5\dfrac{19}{20}$

9. $\dfrac{6}{7} \times \dfrac{3}{4} \times \dfrac{2}{3} = \dfrac{6}{7} \times \dfrac{2}{4} \times \dfrac{3}{3} = \dfrac{6}{7} \times \dfrac{1}{2} \times \dfrac{1}{1} = \dfrac{3}{7} \times \dfrac{1}{1} \times \dfrac{1}{1} = \dfrac{3}{7}$

10. $6 \times \dfrac{2}{3} \times 2\dfrac{5}{6} = \dfrac{6}{1} \times \dfrac{2}{3} \times \dfrac{17}{6} = \dfrac{1}{1} \times \dfrac{2}{3} \times \dfrac{17}{1} = \dfrac{34}{3} = 11\dfrac{1}{3}$

11. $2\dfrac{2}{3} \div 1\dfrac{7}{9} = \dfrac{8}{3} \div \dfrac{16}{9} = \dfrac{8}{3} \times \dfrac{9}{16} = \dfrac{1}{3} \times \dfrac{9}{2} = \dfrac{1}{1} \times \dfrac{3}{2} = \dfrac{3}{2}$

12. $6\dfrac{2}{3} \times 1\dfrac{4}{5} = \dfrac{20}{3} \times \dfrac{9}{5} = \dfrac{4}{3} \times \dfrac{9}{1} = \dfrac{4}{1} \times \dfrac{3}{1} = 12$

DECIMALS

Earlier, we stated that whole numbers are expressed in a system of tens, or the decimal system, using the digits from 0 to 9. This system can be extended to fractions by using a period called a *decimal point*. The digits after a decimal point form a *decimal fraction*. Decimal fractions are smaller than 1—for example, .3, .37, .372, and .105. The first position to the right of the decimal point is called the *tenths' place* since the digit in that position tells how many tenths there are. The second digit to the right of the decimal point is in the *hundredths' place*. The third digit to the right of the decimal point is in the *thousandths' place,* and so on.

Example 1

.3 is a decimal fraction that means

$$3 \times \frac{1}{10} = \frac{3}{10}$$

read "three tenths."

Example 2

The decimal fraction of .37 means

$$3 \times \frac{1}{10} + 7 \times \frac{1}{100} = 3 \times \frac{10}{100} + 7 \times \frac{1}{100}$$

$$= \frac{30}{100} + \frac{7}{100} = \frac{37}{100}$$

read "thirty-seven hundredths."

Example 3

The decimal fraction .372 means

$$\frac{300}{1,000} + \frac{70}{1,000} + \frac{2}{1,000} = \frac{372}{1,000}$$

read "three hundred seventy-two thousandths."

Whole numbers have an understood (unwritten) decimal point to the right of the last digit (e.g., 4 = 4.0). Decimal fractions can be combined with whole numbers to make *decimals*—for example, 3.246, 10.85, and 4.7.

Note: Adding zeros to the right of a decimal after the last digit does not change the value of the decimal.

Rounding Off

Sometimes, a decimal is expressed with more digits than desired. As the number of digits to the right of the decimal point increases, the number increases in accuracy, but a high degree of accuracy is not always needed. Then, the number can be "rounded off" to a certain decimal place.

To round off, identify the place to be rounded off. If the digit to the right of it is 0, 1, 2, 3, or 4, the round-off place digit remains the same. If the digit to the right is 5, 6, 7, 8, or 9, add 1 to the round-off place digit.

Example 1

Round off .6384 to the nearest thousandth. The digit in the thousandths' place is 8. The digit to the right in the ten-thousandths' place is 4, so the 8 stays the same. The answer is .638

Example 2

.6386 rounded to the nearest thousandth is .639, rounded to the nearest hundredth is .64, and rounded to the nearest tenth is .6.

After a decimal fraction has been rounded off to a particular decimal place, all the digits to the right of that place will be 0.

Note: Rounding off whole numbers can be done by a similar method. It is less common but is sometimes used to get approximate answers quickly.

Example 3

Round 32,756 to the nearest *hundred*. This means, to find the multiple of 100 that is nearest the given number. The number in the hundreds' place is 7. The number immediately to the right is 5, so 32,756 rounds to 32,800.

Decimals and Fractions

Changing a Decimal to a Fraction

Place the digits to the right of the decimal point over the value of the place in which the last digit appears and reduce, if possible. The whole number remains the same.

Example

Change 2.14 to a fraction or mixed number. Observe that 4 is the last digit and is in the hundredths' place.

$$.14 = \frac{14}{100} = \frac{7}{50}$$

Therefore:

$$2.14 = 2\frac{7}{50}$$

Changing a Fraction to a Decimal

Divide the numerator of the fraction by the denominator. First, put a decimal point followed by zeros to the right of the number in the numerator. Subtract and divide until there is no remainder. The decimal point in the quotient is aligned directly above the decimal point in the dividend.

Example

Change $\frac{3}{8}$ to a decimal.

Divide

$$\begin{array}{r} .375 \\ 8\overline{)3.000} \\ \underline{24} \\ 60 \\ \underline{56} \\ 40 \\ \underline{40} \end{array}$$

When the division does not terminate with a 0 remainder, two courses are possible.

First Method Divide to three decimal places.

Example

Change $\frac{5}{6}$ to a decimal.

$$\begin{array}{r} .833 \\ 6\overline{)5.000} \\ \underline{48} \\ 20 \\ \underline{18} \\ 20 \\ \underline{18} \\ 2 \end{array}$$

The 3 in the quotient will be repeated indefinitely. It is called an *infinite decimal* and is written .833. . . .

Second Method Divide until there are two decimal places in the quotient, and then write the remainder over the divisor.

Example

Change $\frac{5}{6}$ to a decimal.

$$\begin{array}{r} .833 \\ 6\overline{)5.000} \\ \underline{48} \\ 20 \\ \underline{18} \\ 20 \end{array} = .83\frac{1}{3}$$

Addition

Addition of decimals is both commutative and associative. Decimals are simpler to add than fractions. Place the decimals in a column with the decimal points aligned under each other. Add in the usual way. The decimal point of the answer is also aligned under the other decimal points.

Example

$43 + 2.73 + .9 + 3.01 = ?$

$$\begin{array}{r} 43. \\ 2.73 \\ .9 \\ \underline{3.01} \\ 49.64 \end{array}$$

Subtraction

For subtraction, the decimal points must be aligned under each other. Add zeros to the right of the decimal point if desired. Subtract as with whole numbers.

Examples

$$\begin{array}{rrr} 21.567 & 21.567 & 39.00 \\ -9.4 & -9.48 & -17.48 \\ \hline 12.167 & 12.087 & 21.52 \end{array}$$

Multiplication

Multiplication of decimals is commutative and associative:

$$5.39 \times .04 = .04 \times 5.39$$
$$(.7 \times .02) \times .1 = .7 \times (.02 \times .1)$$

Multiply the decimals as if they were whole numbers. The total number of decimal places in the product is the sum of the number of places (to the right of the decimal point) in all of the numbers multiplied.

Example

$8.64 \times .003 = ?$

8.64	2	places to right of decimal point
× .003	+ 3	places to right of decimal point
.02592	5	places to right of decimal point

A zero had to be added to the left of the product before writing the decimal point to ensure that there would be five decimal places in the product.

Note: To multiply a decimal by 10, simply move the decimal point one place to the right; to multiply by 100, move the decimal point two places to the right.

Division

To divide one decimal (the dividend) by another (the divisor), move the decimal point in the divisor as many places as necessary to the right to make the divisor a whole number. Then move the decimal point in the dividend (expressed or understood) a corresponding number of places, adding zeros if necessary. Then divide as with whole numbers. The decimal point in the quotient is placed above the decimal point in the dividend after the decimal point has been moved.

Example

Divide 7.6 by .32.

$$.32\overline{)7.60} = 32\overline{)760.00}$$

$$\begin{array}{r} 23.75 \\ \hline 64 \\ \hline 120 \\ 96 \\ \hline 240 \\ 224 \\ \hline 160 \\ 160 \\ \hline \end{array}$$

Note: "Divide 7.6 by .32" can be written as $\dfrac{7.6}{.32}$. If this fraction is multiplied by $\dfrac{100}{100}$, an equivalent fraction is obtained with a whole number in the denominator:

$$\frac{7.6}{.32} \times \frac{100}{100} = \frac{760}{32}$$

Moving the decimal point two places to the right in both the divisor and dividend is equivalent to multiplying each number by 100.

Special Cases

If the dividend has a decimal point and the divisor does not, divide as with whole numbers and place the decimal point of the quotient above the decimal point in the divisor.

If both dividend and divisor are whole numbers but the quotient is a decimal, place a decimal point after the last digit of the dividend and add zeros as necessary to get the required degree of accuracy.

Note: To divide any number by 10, simply move its decimal point (understood to be after the last digit for a whole number) one place to the left; to divide by 100, move the decimal point two places to the left; and so on.

PROPERTIES OF NUMBERS

QUIZ

DECIMAL PROBLEMS

1. Change the following fractions into decimals.

   (A) $\frac{5}{8}$

   (B) $\frac{1}{6}$

2. Change the following decimals into fractions and reduce.

   (A) 2.08
   (B) 13.24

3. Change the following decimals into fractions and reduce.

   (A) 17.56
   (B) 21.002

In the following problems, perform the indicated operations.

4. 31.32 + 3.829
5. 5.746 + 354.34
6. 2.567 − 0.021
7. 3.261 − 2.59
8. 73 − .46
9. 0.7 × 3.1
10. 9.2 × 0.03
11. 5.43 + .154 + 17
12. 0.064 ÷ 0.04
13. 0.033 ÷ 0.11
14. Which of the three decimals .09, .769, .8 is the smallest?

SOLUTIONS

1. (A)
$$\begin{array}{r} 0.625 \\ 8\overline{)5.000} \\ \underline{48} \\ 20 \\ \underline{-16} \\ 40 \end{array}$$

   (B)
$$\begin{array}{r} 0.166... \\ 6\overline{)1.0000} \\ \underline{6} \\ 40 \\ \underline{-36} \\ 40 \end{array}$$

2. (A) $2.08 = 2\dfrac{8}{100} = 2\dfrac{2}{25}$

   (B) $13.24 = 13\dfrac{24}{100} = 13\dfrac{6}{25}$

3. (A) $17.56 = 17\dfrac{56}{100} = 17\dfrac{28}{50} = 17\dfrac{14}{25}$

   (B) $21.002 = 21\dfrac{2}{1,000} = 21\dfrac{1}{500}$

4. $\begin{array}{r} 31.32 \\ +\ 3.829 \\ \hline 35.149 \end{array}$

5. $\begin{array}{r} 5.746 \\ +\ 354.34 \\ \hline 360.086 \end{array}$

6. $\begin{array}{r} 2.567 \\ -\ 0.021 \\ \hline 2.546 \end{array}$

7. $\begin{array}{r} 3.261 \\ -\ 2.59 \\ \hline 0.671 \end{array}$

8. $\begin{array}{r} 73.00 \\ -\ .46 \\ \hline 72.54 \end{array}$

9. $\begin{array}{r} 3.1 \\ \times\ 0.7 \\ \hline 2.17 \end{array}$

PROPERTIES OF NUMBERS

10. 9.2   (One digit to the right of the decimal point)
× .003   (Three digits to the right of the decimal point)
.0276   (Four digits to the right of the decimal point)

11.
5.43
.154
+ 17.000
22.584

12. $.04 \overline{)0.064}$ = 1.6

13. $.11 \overline{).033}$ = 0.3

14. The easiest way to determine the smallest decimal number is to append 0s to the end of each of the numbers until they all have the same number of digits. Then, ignore the decimal points and see which number is the smallest. Thus, .09 = .090, .769 = .769, .8 = .800. Clearly, the smallest number is .09.

Percents

Percents, like fractions and decimals, are ways of expressing parts of whole numbers, as 93 percent, 50 percent, and 22.4 percent. Percents are expressions of hundredths—that is, of fractions whose denominator is 100. The symbol for percent is %.

Example

$$25\% = \text{twenty-five hundredths} = \frac{25}{100} = \frac{1}{4}$$

The word *percent* means *per hundred*. Its main use is in comparing fractions with equal denominators of 100.

Relationship with Fractions and Decimals

Changing Percent into Decimal

Divide the percent by 100 and drop the symbol for percent. Add zeros to the left when necessary:

$$30\% = .30 \qquad 1\% = .01$$

Remember that the short method of dividing by 100 is to move the decimal point two places to the left.

PROPERTIES OF NUMBERS

QUIZ

PERCENT PROBLEMS

1. Change the following decimals into percents.
   (A) 0.374
   (B) 13.02

2. Change the following percents into decimals.
   (A) 62.9%
   (B) 0.002%

3. Change the following fractions into percents.
   (A) $\frac{5}{8}$
   (B) $\frac{44}{400}$

4. Change the following percents into fractions.
   (A) 37.5%
   (B) 0.04%

5. Change $12\frac{1}{4}$% to a decimal.

6. Write .07% as both a decimal and a fraction.

7. Write $\frac{11}{16}$ as both a decimal and a percent.

8. Write 1.25 as both a percent and a fraction.

9. Which of the following is the largest: $\frac{5}{8}$, 62%, .628?

SOLUTIONS

1. (A) $0.374 = 37.4\%$
   (B) $13.02 = 1,302\%$

2. (A) $62.9\% = 0.629$
   (B) $00.002\% = 0.00002$

3. (A) $\dfrac{5}{8} = 8\overline{)5.000}^{\,0.625} = 62.5\%$

   (B) $\dfrac{44}{400} = 400\overline{)44.00}^{\,0.11} = 11\%$

4. (A) $37.5\% = 0.375 = \dfrac{375}{1,000} = \dfrac{3}{8}$
   (B) $00.04\% = 0.0004 = \dfrac{4}{10,000} = \dfrac{1}{2,500}$

5. $12\dfrac{1}{4}\% = 12.25\% = 0.1225$

6. $.07\% = 0.0007 = \dfrac{7}{10,000}$

7. $\dfrac{11}{16} = 16\overline{)11.0000}^{\,.6875} = 68.75\%$

8. $1.25 = 125\% = \dfrac{125}{100} = \dfrac{5}{4} = 1\dfrac{1}{4}$

9. In order to determine the largest number, we must write them all in the same form. Writing $\dfrac{5}{8}$ as a decimal, we obtain .625. If we write 62% as a decimal, we get .62. Thus, .628 is the largest of the three numbers.

Solving Percent Word Problems

There are several different types of word problems involving percents that might appear on your test. In addition to generic percent problems, other applications you might be asked to solve involve taxation, commission, profit and loss, discount, and interest. All of these problems are solved in essentially the same way, as the following examples illustrate.

Note that when solving percent problems, it is often easier to change the percent to a decimal or a fraction before computing. When we take a percent of a certain number, that number is called the *base*, the percent we take is called the *rate*, and the result is called the *part*. If we let B represent the base, R represent the rate, and P represent the part, the relationship between these three quantities can be expressed by the following formula:

$$P = R \times B$$

All percent problems can be solved with the help of this formula.

The first four examples below show how to solve all types of generic percent problems. The remaining examples involve specific financial applications.

Example 1

In a class of 24 students, 25% received an A. How many students received an A?

The number of students (24) is the base, and 25% is the rate. Change the rate to a fraction for ease of handling and apply the formula.

$$25\% = \frac{25}{100} = \frac{1}{4}$$

$$P = R \times B$$

$$= \frac{1}{\cancel{4}} \times \frac{\cancel{24}^{6}}{1}$$

$$= 6 \text{ students}$$

To choose between changing the percent (rate) to a decimal or a fraction, simply decide which would be easier to work with. In Example 1, the fraction is easier to work with because cancellation is possible. In Example 2, the situation is the same except for a different rate. This time, the decimal form is easier.

Example 2

In a class of 24 students, 29.17% received an A. How many students received an A? Changing the rate to a fraction yields

$$\frac{29.17}{100} = \frac{2917}{10,000}$$

You can quickly see that the decimal is the better choice.

$$29.17\% = .2917$$
$$P = R \times B$$
$$= .2917 \times 24$$
$$= 7 \text{ students}$$

```
  .2917
× 24
─────
 1.1668
 5.834
─────
 7.0008
```

Example 3

What percent of a 40-hour week is a 16-hour schedule?

40 hours is the base, and 16 hours is the part. $P = R \times B$
$$16 = R \times 40$$

Divide each side of the equation by 40.

$$\frac{16}{40} = R$$

$$\frac{2}{5} = R$$

$$40\% = R$$

Example 4

A woman paid $15,000 as a down payment on a house. If this amount was 20% of the price, what did the house cost?

The part (or percentage) is $15,000, the rate is 20%, and we must find the base. Change the rate to a fraction.

$$20\% = \frac{1}{5}$$

$$P = R \times B$$

$$\$15,000 = \frac{1}{5} \times B$$

Multiply each side of the equation by 5.

$$\$75,000 = B = \text{cost of house}$$

Commission

Example

A salesperson sells a new car for $24,800 and receives a 5% commission. How much commission does he receive?

The cost of the car ($24,800) is the base, and the rate is 5%. We are looking for the amount of commission, which is the part.

$P = 5\% \times \$24,800 = .05 \times \$24,800 = \$1,240$

Thus, the salesperson receives a commission of $1,240.

Taxation

Example

Janet buys a laptop computer for $1,199 and has to pay 7% sales tax. What is the amount of sales tax she owes, and what is the total price of the computer?

The cost of the computer ($1,199) is the base, and the rate is 7%. We are looking for the amount of sales tax, which is the part.

$P = 7\% \times \$1,199 = .07 \times \$1,199 = \$83.93$

Thus, the sales tax is $83.93, and the total cost of the computer is $1,199 + $83.93 = $1,282.93.

Discount

The amount of discount is the difference between the original price and the sale, or discount, price. The rate of discount is usually given as a fraction or as a percent. Use the formula of the percent problems $P = R \times B$, but now P stands for the part or discount, R is the rate, and B, the base, is the original price.

Example 1

A table listed at $160 is marked 20% off. What is the sale price?

$P = R \times B$
$= .20 \times \$160 = \$32$

This is the amount of discount or how much must be subtracted from the original price. Then:

$160 − $32 = $128 sale price

Example 2

A car priced at $9,000 was sold for $7,200. What was the rate of discount?

Amount of discount = $9,000 − $7,200
= $1,800

Discount = rate × original price
$1,800 = R × $9,000

Divide each side of the equation by $9,000:

$$\frac{\cancel{1,800}^{20}}{\cancel{9,000}_{100}} = \frac{20}{100} = R = 20\%$$

Successive Discounting

When an item is discounted more than once, it is called successive discounting.

Example 1

In one store, a dress tagged at $40 was discounted 15%. When it did not sell at the lower price, it was discounted an additional 10%. What was the final selling price?

Discount = R × original price
First discount = .15 × $40 = $6
$40 − $6 = $34 selling price after first discount
Second discount = .10 × $34 = $3.40
$34 − $3.40 = $30.60 final selling price

Example 2

In another store, an identical dress was also tagged at $40. When it did not sell, it was discounted 25% all at once. Is the final selling price lower or higher than in Example 1?

Discount = R × original price
= .25 × $40
= $10
$40 − $10 = $30 final selling price

This is a lower selling price than in Example 1, where two successive discounts were taken. Although the two discounts from Example 1 add up to the discount of Example 2, the final selling price is not the same.

Interest

Interest problems are similar to discount and percent problems. If money is left in the bank for a year and the interest is calculated at the end of the year, the usual formula $P = R \times B$ can be used, where P is the *interest*, R is the *rate*, and B is the *principal* (original amount of money borrowed or loaned).

Example 1

A certain bank pays interest on savings accounts at the rate of 4% per year. If a man has $6,700 on deposit, find the interest earned after 1 year.

$$P = R \times B$$

Interest = rate · principal

$$P = .04 \times \$6,700 = \$268 \text{ interest}$$

Interest problems frequently involve more or less time than 1 year. Then the formula becomes:

Interest = rate × principal × time

Example 2

If the money is left in the bank for 3 years at simple interest (the kind we are discussing), the interest is

3 × $268 = $804

Example 3

Suppose $6,700 is deposited in the bank at 4% interest for 3 months. How much interest is earned?

Interest = rate × principal × time

Here, the 4% rate is for 1 year. Since 3 months is $\frac{3}{12} = \frac{1}{4}$,

Interest = $.04 \times \$6,700 \times \frac{1}{4} = \$67$

Percent of Change Problems

The percent-of-change problem is a special, yet very common, type of percent problem. In such a problem, there is a quantity that has a certain starting value (usually called the "original value"). This original value changes by a certain amount (either an increase or a decrease), leading to what is called the "new value." The problem is to express this increase or decrease as a percent.

Percent-of-change problems are solved by using a method analogous to that used in the above problems. First, calculate the *amount* of the increase or decrease. This amount plays the role of the part P in the formula $P = R \times B$. The base, B, is the original amount, regardless of whether there was a gain or a loss.

Example 1

By what percent does Mary's salary increase if her present salary is $20,000 and she accepts a new job at a salary of $28,000?

Amount of increase is:

$$\$28,000 - \$20,000 = \$8,000$$

$$P = R \times B$$

$$\$8,000 = R \times \$20,000$$

Divide each side of the equation by $20,000. Then:

$$\frac{\cancel{8,000}^{40}}{\cancel{20,000}_{100}} = \frac{40}{100} = R = 40\% \text{ increase}$$

Example 2

On Tuesday, the price of Alpha stock closed at $56 a share. On Wednesday, the stock closed at a price that was $14 higher than the closing price on Tuesday. What was the percent of increase in the closing price of the stock?

In this problem, we are given the amount of increase of $14. Thus,

$$P = R \times B$$

$14 = R \times 56$. Thus,

$$R = \frac{14}{56} = \frac{1}{4} = 25\%.$$

The percent of increase in the closing price of the stock is 25%.

PERCENT WORD PROBLEMS

1. Susan purchased a new refrigerator priced at $675. She made a down payment of 15% of the price. Find the amount of the down payment.

2. After having lunch, Ian leaves a tip of $4.32. If this amount represents 18% of the lunch bill, how much was the bill?

3. Before beginning her diet, Janet weighed 125 pounds. After completing the diet, she weighed 110 pounds. What percent of her weight did she lose?

4. A self-employed individual places $5,000 in an account that earns 8% simple annual interest. How much money will be in this account after 2 years?

5. If a $12,000 car loses 10% of its value every year, what is it worth after 3 years?

6. Peter invests $5,000 at 4% simple annual interest. How much is his investment worth after 2 months?

7. Sales volume at an office supply company climbed from $18,300 last month to $56,730 this month. Find the percent of increase in sales.

8. A men's clothing retailer orders $25,400 worth of outer garments and receives a discount of 15%, followed by an additional discount of 10%. What is the cost of the clothing after these two discounts?

9. Janet receives a 6% commission for selling boxes of greeting cards. If she sells 12 boxes for $40 each, how much does she earn?

10. A small business office bought a used copy machine for 75% of the original price. If the original price was $3,500, how much did the office pay for the copy machine?

11. A lawyer who is currently earning $42,380 annually receives a 6.5% raise. What is his new annual salary?

12. An industrial plant reduces its number of employees, which was originally 3,760, by 5%. How many employees now work at the plant?

SOLUTIONS

1. Amount of down payment = $675 × 15% = $675 × .15
   = $101.25

2. Amount of bill = Amount of tip/Percent of tip = 4.32/0.18
   = $24.

3. Amount of weight lost = 125 − 110 = 15 lb.
   Percent of weight lost = Amount of weight lost/Original weight = 15/125 = 12%

4. Each year, the amount of interest earned is $5,000 × 8% = $400. Thus, in 2 years, $800 in interest is earned, and the account has $5,800 in it.

5. Value of car after 1 year = 12,000 × 0.90 = $10,800
   Value of car after 2 years = 10,800 × 0.90 = $9,720
   Value of car after 3 years = 9,720 × 0.90 = $8,748

6. Value of investment = Principal × Rate × Time
   = 5,000 × 0.04 × 1/6 = $33.33

7. Amount of increase = $56,730 − $18,300 = $38,430
   Percent of increase = 38,430/18,300 = 210%

8. Price after the first markdown = $25,400 × 85% = $21,590
   Price after the second markdown = $21,590 × 90% = $19,431

9. 12 boxes for $40 each cost $480. Since Janet makes a 6% commission, she will receive $480 × 6% = $28.80

10. Cost = 3,500 × 75% = $2,625

11. Amount of raise = 42,380 × 6.5% = $2,754.70
    New Salary = $42,380 + $2,754.70 = $45,134.70

12. Number of employees who lost their jobs = 3,760 × 5% = 188
    Number of employees who now work at the plant
    = 3,760 − 188 = 3,572

SYSTEMS OF MEASUREMENTS

THE ENGLISH SYSTEM

On the SAT II Math Tests, it is sometimes necessary to compute not only in the English system of measurement, but also in the metric system. It may also be necessary to convert from one system to the other, but in such cases, you will be given the appropriate conversion factors. Thus, these factors do not need to be memorized.

As far as the English system is concerned, make sure that you have the following relationships memorized:

Conversion Factors for Length
36 inches = 3 feet = 1 yard
12 inches = 1 foot
5,280 feet = 1,760 yards = 1 mile

Conversion Factors for Volume
2 pints = 1 quart
16 fluid ounces = 1 pint
8 pints = 4 quarts = 1 gallon

Conversion Factors for Weight
16 ounces = 1 pound
2,000 pounds = 1 ton

These conversion factors enable you to change units within the English system.

Examples

1. How many feet are in 5 miles?

    5 miles × (5,280 feet/1 mile) = 26,400 feet

    Notice how the unit of "miles" cancels out of the numerator and denominator.

2. How many ounces are in 2 tons?

    2 tons × (2,000 pounds/1 ton) × (16 ounces/1 pound) = 64,000 ounces

    Notice how the units of "tons" and "pounds" cancel out of the numerator and denominator.

The Metric System

In the metric system, distance or length is measured in meters. Similarly, volume is measured in liters, and mass is measured in grams. The prefixes below are appended to the beginning of these basic units to indicate other units of measure with sizes equal to each basic unit multiplied or divided by powers of 10.

$$\text{giga} = 10^9$$
$$\text{mega} = 10^6$$
$$\text{kilo} = 10^3$$
$$\text{hecto} = 10^2$$
$$\text{deka} = 10^1$$
$$\text{deci} = 10^{-1}$$
$$\text{centi} = 10^{-2}$$
$$\text{milli} = 10^{-3}$$
$$\text{micro} = 10^{-6}$$
$$\text{nano} = 10^{-9}$$
$$\text{pico} = 10^{-12}$$

From the table above, we can see, for example, that a kilometer is 1,000 times as long as a meter, 100,000 times as long as a centimeter, and 1,000,000 times as long as a millimeter. Similarly, a centigram is 1/100 the size of a gram.

Conversions among metric units can be made quickly by moving decimal points.

Examples

1. Convert 9.43 kilometers to meters.

    Since meters are smaller than kilometers, our answer will be larger than 9.43. There are 1,000 meters in a kilometer, so we move the decimal point three places to the right. Therefore, 9.43 kilometers is equal to 9,430 meters.

2. Convert 512 grams to kilograms.

    Since kilograms are more massive than grams, our answer must be less than 512. There are 10^{-3} kilograms in a gram, so we move the decimal point three places to the left. Therefore, 512 grams are equal to .512 kilograms.

SYSTEMS OF MEASUREMENTS

Conversions Between the English and the Metric Systems

Conversions between the English and the metric systems are accomplished in the same way as conversions within the English system. Recall that any problem that requires you to make such a conversion will include the necessary conversion factors.

Examples

1. If 1 meter is equivalent to 1.09 yards, how many yards are in 10 meters?

   10 meters $\times$ (1.09 yards/1 meter) = 10.9 yards.

2. If 1 yard is equivalent to .914 meters, how many meters are there in 24 yards?

   24 yards $\times$ (.914 meters/1 yard) = 21.936 meters.

QUIZ

SYSTEMS OF MEASUREMENT PROBLEMS

1. Express 38 meters in millimeters.

2. Express 871 millimeters in centimeters.

3. Which measurement is greater, 8,000 millimeters or 7 meters?

4. Arrange the following from smallest to largest: 6,700 meters, 672,000 centimeters, and 6.6 kilometers.

5. Express 49 milligrams in centigrams.

6. Express 4.6 liters in milliliters.

7. There are 2.2 pounds in a kilogram. A package weighing 32.5 kilograms is shipped to the U.S. What is its weight in pounds?

8. There are .914 meters in a yard. A line drawn on a blueprint measures 1.5 yards. What is its length in meters?

9. There are .62 miles in a kilometer. If the distance between two exits on a highway is 40 kilometers, what is the distance in miles?

10. There are 1.06 quarts in a liter. A particular brand of bottled water is available in two different bottle sizes—a 2.25 quart bottle and a 2.1 liter bottle. Which bottle contains more water?

SOLUTIONS

1. Since meters are larger than millimeters, our answer will be larger than 38. There are 1,000 millimeters in a meter, so we move the decimal point three places over to the right. Thirty-eight meters is equal to 38,000 millimeters.

2. Since millimeters are smaller than centimeters, our answer will be smaller than 871. There are 10 millimeters in a centimeter, so we move the decimal point one place over to the left. Therefore, 871 millimeters is equal to 87.1 centimeters.

3. In order to answer this question, we must express both measures in the same units. Since, for example, 8,000 millimeters is equal to 8 meters, we can see that 8,000 millimeters is larger than 7 meters.

4. Let's start by expressing all measurements in meters.

   672,000 centimeters = 6,720 meters
   6.6 kilometers = 6,600 meters
   6,700 meters = 6,700 meters

   Thus, from smallest to largest, we have 6.6 kilometers, 6,700 meters, and 672,000 centimeters.

5. Since there are 10 milligrams in a centigram, 49 milligrams are equal to 4.9 centigrams.

6. Since there are 1,000 milliliters in a liter, there are 4,600 milliliters in 4.6 liters.

7. 32.5 kgs = 32.5 kgs $\times$ (2.2 lbs/1 kg) = 71.5 lbs.

8. 1.5 yards = 1.5 yards $\times$ (.914 meters/1 yard) = 1.371 meters.

9. 40 kilometers = 40 kilometers $\times$ (.62 miles/1 kilometer) = 24.8 miles.

10. Express 2.1 liters in quarts.

    2.1 liters = 2.1 liters $\times$ (1.06 quarts/1 liter) = 2.226 quarts. Thus, the quart bottle holds more.

SIGNED NUMBERS

In describing subtraction of whole numbers, we said that the operation was not closed—that is, 4 − 6 will yield a number that is not a member of the set of counting numbers and zero. The set of *integers* was developed to give meaning to such expressions as 4 − 6. The set of integers is the set of all *signed* whole numbers and zero. It is the set {..., −4, −3, −2, −1, 0, 1, 2, 3, 4, ...}

The first three dots symbolize the fact that the negative integers go on indefinitely, just as the positive integers do. Integers preceded by a minus sign (called *negative integers*) appear to the left of 0 on a number line.

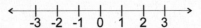

Decimals, fractions, and mixed numbers can also have negative signs. Together with positive fractions and decimals, they appear on the number line in this fashion:

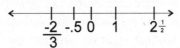

All numbers to the right of 0 are called *positive numbers*. They have the sign +, whether or not it is actually written. Business gains or losses, feet above or below sea level, and temperatures above and below zero can all be expressed by means of signed numbers.

Addition

If the numbers to be added have the same sign, add the numbers (integers, fractions, decimals) as usual and use their common sign in the answer:

$$+9 + (+8) + (+2) = +19 \text{ or } 19$$
$$-4 + (-11) + (-7) + (-1) = -23$$

If the numbers to be added have different signs, add the positive numbers and then the negative numbers. Ignore the signs and subtract the smaller total from the larger total. If the larger total is positive, the answer will be positive; if the larger total is negative, the answer will be negative. The answer may be zero. Zero is neither positive nor negative and has no sign.

Example

$$+3 + (-5) + (-8) + (+2) = ?$$
$$+3 + (+2) = +5$$
$$-5 + (-8) = -13$$
$$13 - 5 = 8$$

Since the larger total (13) has a negative sign, the answer is −8.

SUBTRACTION

The second number in a subtraction problem is called the *subtrahend*. In order to subtract, change the sign of the subtrahend and then continue as if you were *adding* signed numbers. If there is no sign in front of the subtrahend, it is assumed to be positive.

Examples

Subtract the subtrahend (bottom number) from the top number.

```
  15        5      -35     -35      42
   5       15      -42      42      35
  ──       ──      ───     ───      ──
  10      -10        7     -77       7
```

MULTIPLICATION

If only two signed numbers are to be multiplied, multiply the numbers as you would if they were not signed. Then, if the two numbers have the *same sign,* the product is *positive.* If the two numbers have *different signs,* the product is *negative.* If more than two numbers are being multiplied, proceed two at a time in the same way as before, finding the signed product of the first two numbers, then multiplying that product by the next number, and so on. The product has a positive sign if all the factors are positive or there is an even number of negative factors. The product has a negative sign if there is an odd number of negative factors.

Example 1

$-3 \cdot (+5) \cdot (-11) \cdot (-2) = -330$

The answer is negative because there is an odd number (three) of negative factors.

The product of a signed number and zero is zero. The product of a signed number and 1 is the original number. The product of a signed number and −1 is the original number with its sign changed.

Example 2

$-5 \cdot 0 = 0$

$-5 \cdot 1 = -5$

$-5 \cdot (-1) = +5$

DIVISION

If the divisor and the dividend have the same sign, the answer is positive. Divide the numbers as you normally would. If the divisor and the dividend have different signs, the answer is negative. Divide the numbers as you normally would.

Example 1

$$-3 \div (-2) = \frac{3}{2} = 1\frac{1}{2}$$

$$8 \div (-.2) = -40$$

If zero is divided by a signed number, the answer is zero. If a signed number is divided by zero, the answer does not exist. If a signed number is divided by 1, the number remains the same. If a signed number is divided by −1, the quotient is the original number with its sign changed.

Example 2

$$0 \div (-2) = 0$$

$$-\frac{4}{3} \div 0 \quad \text{is not defined}$$

$$\frac{2}{3} \div 1 = \frac{2}{3}$$

$$4 \div -1 = -4$$

QUIZ

SIGNED NUMBERS PROBLEMS

Perform the indicated operations:

1. $+12 + (-10) + (+2) + (-6) =$
2. $+7 + (-2) + (-8) + (+3) =$
3. $-3 - (-7) - (+4) + (-2) =$
4. $-(-8) - 10 - (+12) + (-4) =$
5. $-6 \times (+2) \times (-1) \times (-7) =$
6. $-2 \times (-3) \times (+4) \times (-2) =$
7. $15 \div (-0.5) =$
8. $\dfrac{(+12 \times -2)}{(-4)} =$
9. $(3)(2)(1)(0)(-1)(-2)(-3) =$
10. $\dfrac{(-8)(+3)}{(-6)(-2)(5)} =$
11. $\dfrac{6}{15} \div \left(\dfrac{-12}{5}\right) =$
12. $\dfrac{(+5) - (-13)}{(-4) + (-5)} =$

SOLUTIONS

1. $+12 + (-10) + (+2) + (-6) = +2 + (+2) + (-6)$
   $= 4 + (-6) = -2$

2. $+7 + (-2) = +7 - 2 = +5$
   $+5 + (-8) = +5 - 8 = -3$
   $-3 + (+3) = 0$

3. $-3 - (-7) = -3 + 7 = +4$
   $+4 - (+4) = +4 - 4 = 0$
   $0 + (-2) = -2$

4. $-(-8) - 10 - (+12) + (-4) = +8 - 10 - (+12) + (-4)$
   $= -2 - (+12) + (-4) = -14 + (-4) = -18$

5. $-6 \times (+2) = -12$
   $-12 \times (-1) = +12$
   $+12 \times (-7) = -84$

6. $-2 \times (-3) \times (+4) \times (-2) = +6 \times (+4) \times (-2)$
   $= 24 \times (-2) = 48$

7. $15 \div (-0.5) = -30$

8. $(+12 \times -2) = -24$
   $\dfrac{(-24)}{(-4)} = +6$

9. $(3)(2)(1)(0)(-1)(-2)(-3) = 0$, since, if 0 is a factor in any multiplication, the result is 0.

10. $\dfrac{(-8)(+3)}{(-6)(-2)(5)} = \dfrac{-24}{60} = -\dfrac{2}{5}$

11. $\dfrac{6}{15} \div \left(\dfrac{-12}{5}\right) = \dfrac{6}{15} \times \dfrac{5}{-12} = \dfrac{1}{3} \times \dfrac{1}{-2} = -\dfrac{1}{6}$

12. $\dfrac{(+5) - (-13)}{(-4) + (-5)} = \dfrac{5 + 13}{-9} = \dfrac{18}{-9} = -2$

POWERS, EXPONENTS, AND ROOTS

Exponents

The product $10 \times 10 \times 10$ can be written 10^3. We say 10 is raised to the *third power*. In general, $a \times a \times a \times \ldots \times a$ n times is written a^n. The *base* a is raised to the nth power, and n is called the *exponent*.

Example 1

$3^2 = 3 \times 3$ read "3 squared"
$2^3 = 2 \times 2 \times 2$ read "2 cubed"
$5^4 = 5 \times 5 \times 5 \times 5$ read "5 to the fourth power"

If the exponent is 1, it is usually understood and not written; thus, $a^1 = a$.

Since

$a^2 = a \times a$ and $a^3 = a \times a \times a$

then

$a^2 \times a^3 = (a \times a)(a \times a \times a) = a^5$

There are three rules for exponents. In general, if k and m are any counting numbers or zero, and a is any number,

Rule 1: $a^k \times a^m = a^{k+m}$

Rule 2: $a^m \times b^m = (ab)^m$

Rule 3: $(a^k)^n = a^{kn}$

Example 2

Rule 1: $2^2 \times 2^3 = 4 \times 8 = 32$
and
$2^2 \times 2^3 = 2^5 = 32$

Rule 2: $3^2 \times 4^2 = 9 \times 16 = 144$
and
$3^2 \times 4^2 = (3 \times 4)^2 = 12^2 = 144$

Rule 3: $(3^2)^3 = 9^3 = 729$
and
$(3^2)^3 = 3^6 = 729$

Roots

The definition of roots is based on exponents. If $a^n = c$, where a is the base and n the exponent, a is called the nth *root* of c. This is written $a = \sqrt[n]{c}$. The symbol $\sqrt{\ }$ is called a *radical sign*. Since $5^4 = 625$, $\sqrt[4]{625} = 5$ and 5 is the fourth root of 625. The most frequently used roots are the second (called the square) root and the third (called the cube) root. The square root is written $\sqrt{\ }$ and the cube root is written $\sqrt[3]{\ }$.

Square Roots

If c is a positive number, there are two values, one negative and one positive, which, when multiplied, will produce c.

Example

$+4 \times (+4) = 16$ and $-4 \times (-4) = 16$

The positive square root of a positive number c is called the *principal* square root of c (briefly, the *square root* of c) and is denoted by $\sqrt{c}$:

$$\sqrt{144} = 12$$

If $c = 0$, there is only one square root, 0. If c is a negative number, there is no real number that is the square root of c:

$\sqrt{-4}$ is not a real number

Cube Roots

Both positive and negative numbers have real cube roots. The cube root of 0 is 0. The cube root of a positive number is positive; that of a negative number is negative.

Example

$2 \times 2 \times 2 = 8$

Therefore, $\sqrt[3]{8} = 2 - 3 \times (-3) \times (-3) = -27$

Therefore, $\sqrt[3]{-27} = -3$

Each number has only one real cube root.

Expanded Form

We previously have seen how to write whole numbers in expanded form. Recall, for example, that the number 1,987 can be written as

$$1,987 = 1(1,000) + 9(100) + 8(10) + 7$$

Thus, 1,987 represents a number containing 7 "ones," 8 "tens," 9 "hundreds," and 1 "thousand." Using exponential notation, 1,987 can be written somewhat more compactly as

$$1,987 = 1(10^3) + 9(10^2) + 8(10^1) + 7$$

Example 1

Write the number 50,127 in expanded form using exponential notation.

$$50,127 = 5(10^4) + 0(10^3) + 1(10^2) + 2(10^1) + 7$$

Example 2

What number is represented by the expanded form $7(10^5) + 3(10^3) + 2(10^2) + 5(10^1) + 4$?

Note that there is no term corresponding to 10^4. Thus, the answer is 703,254.

Simplification of Square Roots

Certain square roots can be written in a simplified or reduced form. Just as all fractions should be simplified if possible, all square roots should also be simplified if possible. To simplify a square root means to remove any perfect square factors from under the square root sign.

The simplification of square roots is based on the *Product Rule for Square Roots:*

$$\sqrt{a \times b} = \sqrt{a} \times \sqrt{b}.$$

To illustrate the technique, let us simplify $\sqrt{12}$. Begin by writing 12 as 4×3, thus transforming the number under the square root sign into a product containing the perfect square factor 4.

$$\sqrt{12} = \sqrt{4 \times 3}$$

Then, using the Product Rule, write the square root of the product as the product of the square root.

$$\sqrt{12} = \sqrt{4 \times 3} = \sqrt{4} \times \sqrt{3}$$

Finally, compute $\sqrt{4}$ to obtain the simplified form.

$$\sqrt{12} = \sqrt{4 \times 3} = \sqrt{4} \times \sqrt{3} = 2\sqrt{3}$$

Example 1

Simplify $\sqrt{98}$.

$$\sqrt{98} = \sqrt{2 \times 49}$$
$$= \sqrt{2} \times \sqrt{49}, \quad \text{where 49 is a square number}$$
$$= \sqrt{2} \times 7$$

Therefore, $\sqrt{98} = 7\sqrt{2}$, and the process terminates because there is no whole number whose square is 2. We call $7\sqrt{2}$ a radical expression or simply a *radical*.

Example 2

Which is larger, $(\sqrt{96})^2$ or $\sqrt{2^{14}}$?

$$(\sqrt{96})^2 = \sqrt{96} \times \sqrt{96} = \sqrt{96 \times 96} = 96$$

$\sqrt{2^{14}} = 2^7 = 128$ because $2^{14} = 2^7 \times 2^7$ by Rule 1

or because $\sqrt{2^{14}} = (2^{14})^{1/2} = 2^7$ by Rule 3

Since $128 > 96$, $\sqrt{2^{14}} > (\sqrt{96})^2$

Example 3

Which is larger, $2\sqrt{75}$ or $6\sqrt{12}$?

These numbers can be compared if the same number appears under the radical sign. The greater number is the one with the larger number in front of the radical sign.

$$\sqrt{75} = \sqrt{25 \times 3} = \sqrt{25} \times \sqrt{3} = 5\sqrt{3}$$

Therefore:

$$2\sqrt{75} = 2(5\sqrt{3}) = 10\sqrt{3}$$
$$\sqrt{12} = \sqrt{4 \times 3} = \sqrt{4} \times \sqrt{3} = 2\sqrt{3}$$

Therefore:

$$6\sqrt{12} = 6(2\sqrt{3}) = 12\sqrt{3}$$

Since $12\sqrt{3} > 10\sqrt{3}$, $6\sqrt{12} > 2\sqrt{75}$

Radicals can be added and subtracted only if they have the same number under the radical sign. Otherwise, they must be reduced to expressions having the same number under the radical sign.

Example 4

Add $2\sqrt{18} + 4\sqrt{8} - \sqrt{2}$.

$\sqrt{18} = \sqrt{9 \times 2} = \sqrt{9} \times \sqrt{2} = 3\sqrt{2}$

Therefore:

$2\sqrt{18} = 2(3\sqrt{2}) = 6\sqrt{2}$

and

$\sqrt{8} = \sqrt{4 \times 2} = \sqrt{4} \times \sqrt{2} = 2\sqrt{2}$

Therefore:

$4\sqrt{8} = 4(2\sqrt{2}) = 8\sqrt{2}$

giving

$2\sqrt{18} + 4\sqrt{8} - \sqrt{2} = 6\sqrt{2} + 8\sqrt{2} - \sqrt{2} = 13\sqrt{2}$

Radicals are multiplied using the rule that

$\sqrt[k]{a \times b} = \sqrt[k]{a} \times \sqrt[k]{b}$

Example 5

$\sqrt{2}\left(\sqrt{2} - 5\sqrt{3}\right) = \sqrt{4} - 5\sqrt{6} = 2 - 5\sqrt{6}$

A quotient rule for radicals similar to the Product Rule is:

$\sqrt[k]{\dfrac{a}{b}} = \dfrac{\sqrt[k]{a}}{\sqrt[k]{b}}$

Example 6

$\sqrt{\dfrac{9}{4}} = \dfrac{\sqrt{9}}{\sqrt{4}} = \dfrac{3}{2}$

QUIZ

Exponents, Powers, and Roots Problems

1. Simplify $\sqrt{180}$
2. Find the sum of $\sqrt{45} + \sqrt{125}$
3. Combine $2\sqrt{20} + 6\sqrt{45} - \sqrt{125}$
4. Find the difference of $8\sqrt{12} - 2\sqrt{27}$
5. Simplify $(3\sqrt{32})(7\sqrt{2})$
6. Simplify $\sqrt{7}(2\sqrt{7} - \sqrt{3})$
7. Simplify $\dfrac{(20\sqrt{96})}{5\sqrt{4}}$
8. Divide and simplify $\dfrac{12\sqrt{75}}{4\sqrt{108}}$
9. Evaluate $-3^2 + (3^2)^3$
10. $3^2 \times 2^4 =$
11. Simplify $(\sqrt{15})^2$
12. Simplify $\sqrt{6}\sqrt{3}\sqrt{2}$

SOLUTIONS

1. $\sqrt{180} = \sqrt{36 \times 5} = 6\sqrt{5}$

2. $\sqrt{45} + \sqrt{125} = 3\sqrt{5} + 5\sqrt{5} = 8\sqrt{5}$

3. $2\sqrt{20} + 6\sqrt{45} - \sqrt{125}$
   $= 2\sqrt{4 \times 5} + 6\sqrt{9 \times 5} - \sqrt{25 \times 5}$
   $= 4\sqrt{5} + 18\sqrt{5} - 5\sqrt{5}$
   $= 17\sqrt{5}$

4. $8\sqrt{12} - 2\sqrt{27} = 8\sqrt{4 \times 3} - 2\sqrt{9 \times 3}$
   $= 16\sqrt{3} - \sqrt{3}$ [handwritten note: 6]
   $= 10\sqrt{3}$ [handwritten note: 15]

5. $(3\sqrt{32})(7\sqrt{2}) = 21(\sqrt{64}) = 21(8) = 168$

6. $\sqrt{7}(2\sqrt{7} - \sqrt{3}) = 2\sqrt{7}\sqrt{7} - \sqrt{7}\sqrt{3}$
   $= 2\sqrt{49} - \sqrt{21}$
   $= 2(7) - \sqrt{21}$
   $= 14 - \sqrt{21}$

7. $\dfrac{(20\sqrt{96})}{5\sqrt{4}} = \left(\dfrac{20}{5}\right)\left(\dfrac{\sqrt{96}}{\sqrt{4}}\right) = 4\sqrt{\dfrac{96}{4}}$
   $= 4\sqrt{24}$
   $= 4\sqrt{4 \cdot 6}$
   $= 8\sqrt{6}$

8. $\dfrac{12\sqrt{75}}{4\sqrt{108}} = \dfrac{12\sqrt{25 \times 3}}{4\sqrt{36 \times 3}} = \dfrac{12 \times 5\sqrt{3}}{4 \times 6\sqrt{3}} = \dfrac{60\sqrt{3}}{24\sqrt{3}} = \dfrac{60}{24} = \dfrac{5}{2}$

9. $-3^2 + (3^2)^3 = -9 + 3^6 = -9 + 729 = 720$

10. $3^2 \times 2^4 = 9 \times 16 = 144$

11. $(\sqrt{15})^2 = 15$, since squares and square roots are inverse operations

12. $\sqrt{6}\sqrt{3}\sqrt{2} = \sqrt{6 \times 3 \times 2} = \sqrt{36} = 6$

ALGEBRA

Those who take both the Level IC and the Level IIC tests must know all of the algebra topics contained in the following sections. There are a number of more advanced algebraic topics that are tested on the Level IIC test, and these topics are covered in later sections.

Algebra is a generalization of arithmetic. It provides methods for solving problems that cannot be done by arithmetic alone or that can be done by arithmetic only after long computations. Algebra provides a shorthand way of reducing long verbal statements to brief formulas, expressions, or equations. After the verbal statements have been reduced, the resulting algebraic expressions can be simplified.

Suppose that a room is 12 feet wide and 20 feet long. Its perimeter (measurement around the outside) can be expressed as:

$12 + 20 + 12 + 20$ or $2(12 + 20)$

If the width of the room remains 12 feet but the letter l is used to symbolize length, the perimeter is:

$12 + l + 12 + l$ or $2(12 + l)$

Further, if w is used for width, the perimeter of *any* rectangular room can be written as $2(w + l)$. This same room has an area of 12 feet by 20 feet, or 12×20. If l is substituted for 20, any room of width 12 has area equal to $12l$. If w is substituted for the number 12, the area of any rectangular room is given by wl or lw. Expressions such as wl and $2(w + l)$ are called *algebraic expressions*. An *equation* is a statement that two algebraic expressions are equal. A *formula* is a special type of equation.

EVALUATING FORMULAS

If we are given an expression and numerical values to be assigned to each letter, the expression can be evaluated.

Example 1

Evaluate $2x + 3y - 7$ if $x = 2$ and $y = -4$.

Substitute given values.

$2(2) + 3(-4) - 7 = ?$

Multiply numbers using rules for signed numbers.

$4 + -12 - 7 = ?$

Collect numbers.

$4 - 19 = -15$

We have already evaluated formulas in arithmetic when solving percent, discount, and interest problems.

Example 2

Evaluate each of the following expressions if $a = 3$, $b = -2$, and $c = 0$.

    a. $-a^2$
    b. $3b - 4b^2$
    c. $ab + 4c$

a. $-a^2 = -(3)^2 = -(9) = -9$

b. $b - 4b^2 = 3(-2) - 4(-2)^2 = -6 - 4(4) = -6 - 16 = -22$

c. $ab + 4c = (3)(-2) + 4(0) = -6 + 0 = -6$

Example 3

If $x = 1$ and $y = -2$, find the value of $-x^2y^2$

$-x^2y^2 = -(1)^2(-2)^2 = -(1)(4) = -(4) = -4$

Example 4

The formula for temperature conversion is

$$F = \frac{9}{5}C + 32$$

where C stands for the temperature in degrees Celsius and F for degrees Fahrenheit. Find the Fahrenheit temperature that is equivalent to 20°C.

$$F = \frac{9}{5}(20°C) + 32 = 36 + 32 = 68°F$$

Example 5

The formula for the area of a triangle is $A = \dfrac{bh}{2}$. Find A if $b = 12$ and $h = 7$.

$$A = \frac{bh}{2} = \frac{12 \times 7}{2} = 42$$

Algebraic Expressions

Formulation

A more difficult problem than evaluating an expression or formula is translating from a verbal expression to an algebraic one:

Verbal	Algebraic
Thirteen more than x	$x + 13$
Six less than twice x	$2x - 6$
The square of the sum of x and 5	$(x + 5)^2$
The sum of the square of x and the square of 5	$x^2 + 5^2$
The distance traveled by a car going 50 miles an hour for x hours	$50x$
The average of 70, 80, 85, and x	$\dfrac{70 + 80 + 85 + x}{4}$

Simplification

After algebraic expressions have been formulated, they can usually be simplified by means of the laws of exponents and the common operations of addition, subtraction, multiplication, and division. These techniques are described in the next section. Algebraic expressions and equations frequently contain parentheses, which are removed in the process of simplifying. If an expression contains more than one set of parentheses, remove the inner set first and then the outer set. Brackets, [ ], which are often used instead of parentheses, are treated the same way. Parentheses are used to indicate multiplication. Thus, $3(x + y)$ means that 3 is to be multiplied by the sum of x and y. The *distributive law* is used to accomplish this:

$$a(b + c) = ab + ac$$

The expression in front of the parentheses is multiplied by each term inside. Rules for signed numbers apply.

Example 1

Simplify $3[4(2 - 8) - 5(4 + 2)]$.

This can be done in two ways.

Method 1: Combine the numbers inside the parentheses first:

$$3[4(2 - 8) - 5(4 + 2)] = 3[4(-6) - 5(6)]$$
$$= 3[-24 - 30]$$
$$= 3[-54] = -162$$

Method 2: Use the distributive law:

$$3[4(2 - 8) - 5(4 + 2)] = 3[8 - 32 - 20 - 10]$$
$$= 3[8 - 62]$$
$$= 3[-54] = -162$$

MATHEMATICS IC AND IIC REVIEW

If there is a (+) before the parentheses, the signs of the terms inside the parentheses remain the same when the parentheses are removed. If there is a (−) before the parentheses, the sign of each term inside the parentheses changes when the parentheses are removed.

Once parentheses have been removed, the order of operations is multiplication and division, then addition and subtraction from left to right.

Example 2

$(-15 + 17) \times 3 - [(4 \times 9) \div 6] = ?$

Work inside the parentheses first: $(2) \times 3 - [36 \div 6] = ?$

Then work inside the brackets: $2 \times 3 - [6] = ?$

Multiply first, then subtract, proceeding from left to right: $6 - 6 = 0$

The placement of parentheses and brackets is important. Using the same numbers as above with the parentheses and brackets placed in different positions can give many different answers.

Example 3

$-15 + [(17 \times 3) - (4 \times 9)] \div 6 = ?$

Work inside the parentheses first:
$-15 + [(51) - (36)] \div 6 = ?$

Then work inside the brackets:
$-15 + [15] \div 6 = ?$

Since there are no more parentheses or brackets, proceed from left to right, dividing before adding:

$-15 + 2\frac{1}{2} = -12\frac{1}{2}$

Operations

When letter symbols and numbers are combined with the operations of arithmetic (+, −, ×, ÷) and with certain other mathematical operations, we have an *algebraic expression*. Algebraic expressions are made up of several parts connected by a plus or a minus sign; each part is called a *term*. Terms with the same letter part are called *like terms*. Since algebraic expressions represent numbers, they can be added, subtracted, multiplied, and divided.

When we defined the commutative law of addition in arithmetic by writing $a + b = b + a$, we meant that a and b could represent any number. The expression $a + b = b + a$ is an *identity* because it is true for all numbers. The expression $n + 5 = 14$ is not an identity because it is not true for all numbers; it becomes true only when the number 9 is substituted for n. Letters used to represent numbers are called *variables*. If a number stands alone (the 5 or 14 in $n + 5 = 14$), it is called a *constant* because its value is constant or unchanging. If a number appears in front of a variable, it is called a *coefficient*. Because the letter x is frequently used to represent a variable, or *unknown*, the times sign, ×, which can be confused with it in handwriting, is rarely used to express multiplication in algebra. Other expressions used for multiplication are a dot, parentheses, or simply writing a number and letter together:

$5 \cdot 4$ or $5(4)$ or $5a$

Of course, 54 still means fifty-four.

Addition and Subtraction

Only like terms can be combined. Add or subtract the coefficients of like terms, using the rules for signed numbers.

Example 1

Add $x + 2y - 2x + 3y$.

$x - 2x + 2y + 3y = -x + 5y$

Example 2

Perform the subtraction:

$$\begin{array}{r} -30a - 15b + 4c \\ -(-5a + 3b - c + d) \end{array}$$

Change the sign of each term in the subtrahend and then add, using the rules for signed numbers:

$$\begin{array}{r} -30a - 15b + 4c \\ 5a - 3b + c - d \\ \hline -25a - 18b + 5c - d \end{array}$$

Example 3

Perform the following subtraction:

$(h^2 + 6hk - 7k^2) - (3h^2 + 6hk - 10k^2)$

Once again, change the sign of each term in the subtrahend and then add.

$(h^2 + 6hk - 7k^2) - (3h^2 + 6hk - 10k^2)$
$= (h^2 + 6hk - 7k^2) + (-3h^2 - 6hk + 10k^2)$
$= -2h^2 + 3k^2$

Multiplication

Multiplication is accomplished by using the *distributive property*. If the multiplier has only one term, then
$a(b + c) = ab + bc$

Example 1

$9x(5m + 9q) = (9x)(5m) + (9x)(9q)$

$= 45mx + 81qx$

When the multiplier contains more than one term and you are multiplying two expressions, multiply each term of the first expression by each term of the second and then add like terms. Follow the rules for signed numbers and exponents at all times.

Example 2

$(2x - 1)(x + 6)$
$= 2x(x + 6) - 1(x + 6)$
$= 2x^2 + 12x - x - 6$
$= 2x^2 + 11x - 6$

Example 3

$(3x + 8)(4x^2 + 2x + 1)$
$= 3x(4x^2 + 2x + 1) + 8(4x^2 + 2x + 1)$
$= 12x^3 + 6x^2 + 3x + 32x^2 + 16x + 8$
$= 12x^3 + 38x^2 + 19x + 8$

If more than two expressions are to be multiplied, multiply the first two, then multiply the product by the third factor, and so on, until all factors have been used.

Algebraic expressions can be multiplied by themselves (squared) or raised to any power.

Example 4

$$(a + b)^2 = (a + b)(a + b)$$
$$= a(a + b) + b(a + b)$$
$$= a^2 + ab + ba + b^2$$
$$= a^2 + 2ab + b^2$$

since $ab = ba$ by the commutative law.

Example 5

$$(a + b)(a - b) = a(a - b) + b(a - b)$$
$$= a^2 - ab + ba - b^2$$
$$= a^2 - b^2$$

Factoring

When two or more algebraic expressions are multiplied, each is called a factor and the result is the *product*. The reverse process of finding the factors when given the product is called *factoring*. A product can often be factored in more than one way. Factoring is useful in multiplication, division, and solving equations.

One way to factor an expression is to remove any single-term factor that is common to each of the terms and write it outside the parentheses. It is the distributive law that permits this.

Example 1

$$3x + 12 = 3(x + 4)$$

The result can be checked by multiplication.

Example 2

$$3x^3 + 6x^2 + 9x = 3x(x^2 + 2x + 3)$$

The result can be checked by multiplication.

Expressions containing squares can sometimes be factored into expressions containing letters raised to the first power only, called *linear factors*. We have seen that

$$(a + b)(a - b) = a^2 - b^2$$

Therefore, if we have an expression in the form of a difference of two squares, it can be factored as:

$$a^2 - b^2 = (a + b)(a - b)$$

Example 3

Factor $x^2 - 16$.

$$x^2 - 16 = (x)^2 - (4)^2 = (x - 4)(x + 4)$$

Example 4

Factor $4x^2 - 9$.

$$4x^2 - 9 = (2x)^2 - (3)^2 = (2x + 3)(2x - 3)$$

Again, the result can be checked by multiplication.

A third type of expression that can be factored is one containing three terms, such as $x^2 + 5x + 6$. Since

$$\begin{aligned}(x + a)(x + b) &= x(x + b) + a(x + b) \\ &= x^2 + xb + ax + ab \\ &= x^2 + (a + b)x + ab\end{aligned}$$

an expression in the form $x^2 + (a + b)x + ab$ can be factored into two factors of the form $(x + a)$ and $(x + b)$. We must find two numbers whose product is the constant in the given expression and whose sum is the coefficient of the term containing x.

Example 5

Find factors of $x^2 + 5x + 6$.

First find two numbers that, when multiplied, have +6 as a product. Possibilities are 2 and 3, −2 and −3, 1 and 6, −1 and −6. From these select the one pair whose sum is 5. The pair 2 and 3 is the only possible selection, and so:

$$x^2 + 5x + 6 = (x + 2)(x + 3) \quad \text{written in either order.}$$

Example 6

Factor $x^2 - 5x - 6$.

Possible factors of -6 are -1 and 6, 1 and -6, 2 and -3, -2 and 3. We must select the pair whose sum is -5. The only pair whose sum is -5 is $+1$ and -6, and so

$$x^2 - 5x - 6 = (x + 1)(x - 6)$$

In factoring expressions of this type, notice that if the last sign is plus, both a and b have the same sign and it is the same as the sign of the middle term. If the last sign is minus, the numbers have opposite signs.

Many expressions cannot be factored.

Example 7

Factor $2x^3 - 8x^2 + 8x$.

In expressions of this type, begin by factoring out the largest common monomial factor, then try to factor the resulting trinomial.

$$2x^3 - 8x^2 + 8x = 2x(x^2 - 4x + 4) = 2x(x - 2)(x - 2)$$
$$= 2x(x - 2)^2$$

Division

Method 1

$$\frac{36mx^2}{9m^2x} = 4m^1 x^2 m^{-2} x^{-1}$$

$$= 4m^{-1} x^1 = \frac{4x}{m}$$

Method 2

Cancellation

$$\frac{36mx^2}{9m^2x} = \frac{\overset{4}{\cancel{36}}m\cancel{x}x}{\underset{1}{\cancel{9}m\cancel{x}m}} = \frac{4x}{m}$$

This is acceptable because

$$\frac{ab}{bc} = \frac{a}{b}\left(\frac{c}{c}\right) \text{ and } \frac{c}{c} = 1$$

so that $\dfrac{ac}{bc} = \dfrac{a}{b}$

Example 1

If the divisor contains only one term and the dividend is a sum, divide each term in the dividend by the divisor and simplify as you did in Method 2.

$$\frac{9x^3 + 3x^2 + 6x}{3x} = \frac{9x^3}{3x} + \frac{3x^2}{3x} + \frac{6x}{3x} = 3x^2 + x + 2$$

This method cannot be followed if there are two terms or more in the denominator since

$$\frac{a}{b+c} \neq \frac{a}{b} + \frac{a}{c}$$

In this case, write the example as a fraction. Factor the numerator and denominator if possible. Then use laws of exponents or cancel.

Example 2

Divide $x^3 - 9x$ by $x^3 + 6x^2 + 9x$

Write as:

$$\frac{x^3 - 9x}{x^3 + 6x^2 + 9x}$$

Both numerator and denominator can be factored to give:

$$\frac{x(x^2 - 9)}{x(x^2 + 6x + 9)} = \frac{x(x+3)(x-3)}{x(x+3)(x+3)} = \frac{x-3}{x+3}$$

QUIZ

Algebra Problems

1. Simplify: $3[4(6 - 14) - 8(-2 - 5)]$
2. Simplify: $(5x^2 - 3x + 2) - (3x^2 + 5x - 1) + (6x^2 - 2)$
3. Add: $(a - b - c) + (a - b - c) - (a - b - c)$
4. Multiply: $(x - 2)(x^2 + 3x + 7)$
5. Multiply: $(a + 1)^2 (a + 2)$
6. Multiply: $(2x + 1)(3x^2 - x + 6)$
7. Factor completely: $6x^2 - 3x - 18$
8. Factor completely: $12x^2 + 14x + 4$
9. Factor completely: $6x^4 - 150x^2$
10. Factor completely: $4a^2b + 12ab - 72b$
11. Multiply: $\dfrac{x^2 - x - 6}{x^2 - 9} \times \dfrac{x^2 + 4x + 3}{x^2 + 5x + 6}$
12. Simplify: $\dfrac{x^2 - 4x - 21}{x^2 - 9x + 14}$

SOLUTIONS

1. $3[4(6 - 14) - 8(-2 - 5)] = 3[4(-8) - 8(-7)]$
   $= 3[-32 + 56] = 3(24) = 72$

2. $(5x^2 - 3x + 2) - (3x^2 + 5x - 1) + (6x^2 - 2)$
   $= 5x^2 - 3x + 2 - 3x^2 - 5x + 1 + 6x^2 - 2 = 8x^2 - 8x + 1$

3. $(a - b - c) + (a - b - c) - (a - b - c)$
   $= a - b - c + a - b - c - a + b + c = a - b - c$

4. $(x - 2)(x^2 + 3x + 7) = x(x^2 + 3x + 7) - 2(x^2 + 3x + 7)$
   $= x^3 + 3x^2 + 7x - 2x^2 - 6x - 14 = x^3 + x^2 + x - 14$

5. $(a + 1)^2 (a + 2) = (a + 1)(a + 1)(a + 2)$
   $= (a^2 + 2a + 1)(a + 2)$
   $= a^3 + 2a^2 + 2a^2 + 4a + a + 2$
   $= a^3 + 4a^2 + 5a + 2$

6. $(2x + 1)(3x^2 - x + 6) = 2x(3x^2 - x + 6) + 1(3x^2 - x + 6)$
   $= 6x^3 - 2x^2 + 12x + 3x^2 - x + 6$
   $= 6x^3 + x^2 + 11x + 6$

7. $6x^2 - 3x - 18 = 3(2x^2 - x - 6) = 3(2x + 3)(x - 2)$

8. $12x^2 + 14x + 4 = 2(6x^2 + 7x + 2) = 2(3x + 2)(2x + 1)$

9. $6x^4 - 150x^2 = 6x^2(x^2 - 25) = 6x^2(x - 5)(x + 5)$

10. $4a^2b + 12ab - 72b = 4b(a^2 + 3a - 18)$
    $= 4b(a + 6)(a - 3)$

11. $\dfrac{x^2 - x - 6}{x^2 - 9} \times \dfrac{x^2 + 4x + 3}{x^2 + 5x + 6} = \dfrac{(x - 3)(x + 2)}{(x - 3)(x + 3)} \times \dfrac{(x + 3)(x + 1)}{(x + 2)(x + 3)}$
    $= \dfrac{\cancel{(x - 3)}\cancel{(x + 2)}}{\cancel{(x - 3)}(x + 3)} \times \dfrac{\cancel{(x + 3)}(x + 1)}{\cancel{(x + 2)}\cancel{(x + 3)}}$
    $= \dfrac{x + 1}{x + 3}$

12. $\dfrac{x^2 - 4x - 21}{x^2 - 9x + 14} = \dfrac{(x + 3)(x - 7)}{(x - 7)(x - 2)} = \dfrac{x + 3}{x - 2}$

Solving Equations

Solving equations is one of the major objectives in algebra. If a variable x in an equation is replaced by a value or expression that makes the equation a true statement, the value or expression is called a *solution* of the equation. (Remember that an equation is a mathematical statement that one algebraic expression is equal to another.)

An equation may contain one or more variables. We begin with one variable. Certain rules apply to equations whether there are one or more variables. The following rules are applied to give equivalent equations that are simpler than the original:

Addition: If $s = t$, then $s + c = t + c$.
Subtraction: If $s + c = t + c$, then $s = t$.
Multiplication: If $s = t$, then $cs = ct$.
Division: If $cs = ct$ and $c \neq 0$, then $s = t$.

To solve for x in an equation in the form $ax = b$ with $a \neq 0$, divide each side of the equation by a:

$$\frac{ax}{a} = \frac{b}{a} \quad \text{yielding} \quad x = \frac{b}{a}$$

Then, $\frac{b}{a}$ is the solution to the equation.

Example 1

Solve $x + 5 = 12$

Subtract 5 from both sides.

$$\begin{array}{r} x + 5 = 12 \\ -5 \quad -5 \\ \hline x = 7 \end{array}$$

Example 2

Solve $4x = 8$

Write $\frac{4x}{4} = \frac{8}{4}$

$x = 2$

Example 3

Solve $\frac{x}{4} = 9$

Write $4 \times \frac{x}{4} = 9 \times 4$

Thus, $x = 36$

Example 4

Solve $3x + 7 = 19$.

$3x = 12$ Subtract 7 from both sides.

$x = 4$ Divide each side by 3.

Example 5

Solve $2x - (x - 4) = 5(x + 2)$ for x.

$$2x - (x - 4) = 5(x + 2)$$

$2x - x + 4 = 5x + 10$ Remove parentheses by distributive law.

$x + 4 = 5x + 10$ Combine like terms.

$x = 5x + 6$ Subtract 4 from each side.

$-4x = 6$ Subtract $5x$ from each side.

$x = \dfrac{6}{-4}$ Divide each side by -4.

$= -\dfrac{3}{2}$ Reduce fraction to lowest terms.

Negative sign now applies to the entire fraction.

Check the solution for accuracy by substituting in the original equation:

$$2\left(-\dfrac{3}{2}\right) - \left(-\dfrac{3}{2} - 4\right) \stackrel{?}{=} 5\left(-\dfrac{3}{2} + 2\right)$$

$$-3 - \left(-\dfrac{11}{2}\right) \stackrel{?}{=} 5\left(\dfrac{1}{2}\right)$$

$$-3 + \dfrac{11}{2} \stackrel{?}{=} \dfrac{5}{2}$$

$$-\dfrac{6}{2} + \dfrac{11}{2} \stackrel{?}{=} \dfrac{5}{2} \quad \text{check}$$

ALGEBRA

QUIZ

Equation Problems

Solve the following equations for x:

1. $-5x + 3 = x + 2$
2. $6 + 4x = 6x - 10$
3. $x + 3(2x + 5) = -20$
4. $4(x + 2) - (2x + 1) = x + 5$
5. $3(2x + 5) = 10x + 7 + 2(x - 8)$
6. $3x - 4(3x - 2) = -x$
7. $6 + 8(8 - 2x) = 14 - 8(4x - 2)$
8. $\dfrac{2x + 3}{5} - 10 = \dfrac{4 - 3x}{2}$
9. $3(2x + 1) + 2(3x + 1) = 17$
10. $(x - 5)^2 = 4 + (x + 5)^2$

SOLUTIONS

1. $\begin{aligned} -5x + 3 &= x + 2 \\ +5x & +5x \\ \hline 3 &= 6x + 2 \\ -2 & -2 \\ \hline 1 &= 6x \\ \frac{1}{6} &= x \end{aligned}$

2. $\begin{aligned} 6 + 4x &= 6x - 10 \\ 6 &= 2x - 10 \\ 16 &= 2x \\ x &= 8 \end{aligned}$

3. $\begin{aligned} x + 3(2x + 5) &= -20 \\ x + 6x + 15 &= -20 \\ 7x + 15 &= -20 \\ 7x &= -35 \\ x &= -5 \end{aligned}$

4. $\begin{aligned} 4(x + 2) - (2x + 1) &= x + 5 \\ 4x + 8 - 2x - 1 &= x + 5 \\ 2x + 7 &= x + 5 \\ x &= -2 \end{aligned}$

5. $\begin{aligned} 3(2x + 5) &= 10x + 7 + 2(x - 8) \\ 6x + 15 &= 10x + 7 + 2x - 16 \\ 6x + 15 &= 12x - 9 \\ 24 &= 6x \\ x &= 4 \end{aligned}$

6. $\begin{aligned} 3x - 4(3x - 2) &= -x \\ 3x - 12x + 8 &= -x \\ -9x + 8 &= -x \\ 8 &= 8x \\ x &= 1 \end{aligned}$

7. $6 + 8(8 - 2x) = 14 - 8(4x - 2)$
$6 + 64 - 16x = 14 - 32x + 16$
$70 - 16x = 30 - 32x$
$16x = -40$
$x = -\dfrac{40}{16} = -\dfrac{5}{2}$

8. $\dfrac{2x + 3}{5} - 10 = \dfrac{4 - 3x}{2}$
$10 \times \dfrac{2x + 3}{5} - 10 \times 10 = \dfrac{4 - 3x}{2} \times 10$
$2(2x + 3) - 100 = 5(4 - 3x)$
$4x + 6 - 100 = 20 - 15x$
$4x - 94 = 20 - 15x$
$4x = 114 - 15x$
$19x = 114$
$x = 6$

9. $3(2x + 1) + 2(3x + 1) = 17$
$6x + 3 + 6x + 2 = 17$
$12x + 5 = 17$
$12x = 12$
$x = 1$

10. $(x - 5)^2 = 4 + (x + 5)^2$
$x^2 - 10x + 25 = 4 + x^2 + 10x + 25$

Subtract x^2 from both sides and combine terms.
$-10x + 25 = 10x + 29$
$20x = -4$
$x = -\dfrac{1}{5}$

Word Problems Involving One Unknown

In many cases, if you read a word problem carefully, assign a letter to the quantity to be found, and understand the relationships between known and unknown quantities, you can formulate an equation with one unknown.

Number Problems and Age Problems

These two kinds of problems are similar to one another.

Example 1

One number is 3 times another, and their sum is 48. Find the two numbers.

Let x = second number. Then the first is $3x$. Since their sum is 48,

$3x + x = 48$
$4x = 48$
$x = 12$

Therefore, the first number is $3x = 36$.

$36 + 12 = 48$

Example 2

Art is now three times older than Ryan. Four years ago, Art was five times as old as Ryan was then. How old is Art now?

Let R = Ryan's age.

Then $3R$ = Art's age.

Four years ago, Ryan's age was $R - 4$, and Art's age was $3R - 4$.

Since at that time Art was five times as old as Ryan, we have

$5(R - 4) = 3R - 4$
$5R - 20 = 3R - 4$
$2R = 16$
$R = 8, 3R = 24.$

Art is 24 years old now.

Distance Problems

The basic concept is:

Distance = rate · time

Example 1

In a mileage test, a man drives a truck at a fixed rate of speed for 1 hour. Then he increases the speed by 20 miles per hour and drives at that rate for 2 hours. He then reduces that speed by 5 miles per hour and drives at that rate for 3 hours. If the distance traveled was 295 miles, what are the rates of speed over each part of the test?

Let x be the first speed, $x + 20$ the second, and $x + (20 - 5) = x + 15$ the third. Because distance = rate · time, multiply these rates by the time and formulate the equation by separating the two equal expressions for distance by an equal sign:

$$1x + 2(x + 20) + 3(x + 15) = 295$$
$$x + 2x + 3x + 40 + 45 = 295$$
$$6x = 210$$
$$x = 35$$

The speeds are 35, 55, and 50 miles per hour.

Example 2

Two trains leave the Newark station at the same time traveling in opposite directions. One travels at a rate of 60 mph, and the other travels at a rate of 50 mph. In how many hours will the trains be 880 miles apart?

The two trains will be 880 miles apart when the sum of the distances that they both have traveled is 880 miles.

Let r_1 = the rate of the first train; r_2 = the rate of the second train.

Let t_1 = the time of the first train; t_2 = the time of the second train.

Then, the distance the first train travels is $r_1 t_1$, and the distance the second train travels is $r_2 t_2$. Our equation will be $r_1 t_1 + r_2 t_2 = 880$. Since $r_1 = 60$, $r_2 = 50$, and $t_1 = t_2$, we can rewrite the equation as

$$60t + 50t = 880$$
$$110t = 880$$
$$t = 8$$

It will take 8 hours for the trains to get 880 miles apart.

Consecutive Number Problems

This type of problem usually involves only one unknown. Two numbers are consecutive if one is the successor of the other. Three consecutive numbers are of the form x, $x + 1$, and $x + 2$. Since an even number is divisible by 2, consecutive even numbers are of the form $2x$, $2x + 2$, and $2x + 4$. An odd number is of the form $2x + 1$.

Example 1

Find three consecutive whole numbers whose sum is 75.

Let the first number be x, the second $x + 1$, and the third $x + 2$. Then:

$$x + (x + 1) + (x + 2) = 75$$
$$3x + 3 = 75$$
$$3x = 72$$
$$x = 24$$

The numbers whose sum is 75 are 24, 25, and 26. Many versions of this problem have no solution. For example, no three consecutive whole numbers have a sum of 74.

Example 2

Find three consecutive even integers whose sum is 48.

We can express three consecutive even integers as x, $x + 2$, and $x + 4$. Thus, we have:

$$x + (x + 2) + (x + 4) = 48$$
$$3x + 6 = 48$$
$$3x = 42$$
$$x = 14$$

The integers are 14, 16, and 18.

Work Problems

These problems concern the speed with which work can be accomplished and the time necessary to perform a task if the size of the work force is changed.

Example 1

If Joe can type a chapter alone in 6 days and Ann can type the same chapter in 8 days, how long will it take them to type the chapter if they both work on it?

We let x = number of days required if they work together and then put our information into tabular form:

	Joe	Ann	Together
Days to type chapter	6	8	x
Part typed in 1 day	$\frac{1}{6}$	$\frac{1}{8}$	$\frac{1}{x}$

Since the part done by Joe in 1 day plus the part done by Ann in 1 day equals the part done by both in 1 day, we have:

$$\frac{1}{6} + \frac{1}{8} = \frac{1}{x}$$

Next we multiply each member by $48x$ to clear the fractions, giving:

$8x + 6x = 48$

$14x = 48$

$x = 3\frac{3}{7}$ days

Example 2

Working alone, one pipe can fill a pool in 8 hours, a second pipe can fill the pool in 12 hours, and a third can fill it in 24 hours. How long would it take all three pipes, working at the same time, to fill the pool?

Using the same logic as in the previous problem, we obtain the equation

$$\frac{1}{8} + \frac{1}{12} + \frac{1}{24} = \frac{1}{x}$$

To clear the fractions, we multiply each side by $24x$, giving:

$3x + 2x + x = 24$

$6x = 24$

$x = 4$

It would take the pipes 4 hours to fill the pool.

QUIZ

WORD PROBLEMS INVOLVING ONE UNKNOWN

1. One integer is two more than a second integer. The first integer added to four times the second is equal to 17. Find the values of the two integers.

2. If 6 times a number is decreased by 4, the result is the same as when 3 times the number is increased by 2. What is the number?

3. The smaller of two numbers is 31 less than three times the larger. If the numbers differ by 7, what is the smaller number?

4. The sum of three consecutive even integers is 84. Find the smallest of the integers.

5. In a recent local election with two candidates, the winner received 372 more votes than the loser. If the total number of votes cast was 1,370, how many votes did the winning candidate receive?

6. Mike is three years older than Al. In nine years, the sum of their ages will be 47. How old is Mike now?

7. At the Wardlaw Hartridge School Christmas program, student tickets cost $3, and adult tickets cost twice as much. If a total of 200 tickets were sold, and $900 was collected, how many student tickets were sold?

8. Mrs. Krauser invested a part of her $6,000 inheritance at 9 percent simple annual interest and the rest at 12 percent simple annual interest. If the total interest earned in one year was $660, how much did she invest at 12 percent?

9. One pump working continuously can fill a reservoir in thirty days. A second pump can fill the reservoir in twenty days. How long would it take both pumps working together to fill the reservoir?

10. Working together, Brian, Peter, and Jared can shovel the driveway in 12 minutes. If Brian, alone, can shovel the driveway in 21 minutes, and Peter, alone, can shovel the driveway in 84 minutes, how long would it take Jared to shovel the driveway alone?

11. Jimmy is now three years older than Bobby. If seven years from now the sum of their ages is 79, how old is Jimmy now?

12. A freight train and a passenger train leave the same station at noon and travel in opposite directions. If the freight train travels 52 mph and the passenger train travels 84 mph, at what time are they 680 miles apart?

ALGEBRA

SOLUTIONS

1. Let x = the first integer. Then,
   $x - 2$ = the second integer, and
   $$x + 4(x - 2) = 17$$
   $$x + 4x - 8 = 17$$
   $$5x - 8 = 17$$
   $$5x = 25$$
   $$x = 5$$

   The second integer is $x - 2 = 5 - 2 = 3$.

   The integers are 3 and 5.

2. Let x = the number. Then,
   $6x - 4 = 3x + 2$ Thus,
   $3x = 6$ And
   $x = 2$ The number is 2.

3. Let S = the smaller number.
   Then, the larger number = $S + 7$, and
   $$S + 31 = 3(S + 7)$$
   $$S + 31 = 3S + 21$$
   $$2S = 10$$
   $$S = 5 \quad \text{The smaller number is 5.}$$

4. Let x = the smallest integer. Then,
   $x + 2$ = the middle integer, and
   $x + 4$ = the largest integer
   $$x + (x + 2) + (x + 4) = 84$$
   $$3x + 6 = 84$$
   $$3x = 78$$
   $$x = 26$$

   The smallest of the three integers is 26.

5. Let W = the number of votes the winner received.
   Then, $W - 372$ = the number of votes the loser received.
   $$W + (W - 372) = 1,370$$
   $$2W - 372 = 1,370$$
   $$2W = 1,742$$
   $$W = 871$$

   The winner received 871 votes.

6. Let M = Mike's age now. Then, $M - 3$ = Al's age. In nine years, Mike will be $M + 9$, and Al will be $M + 6$. Therefore,

$$M + 9 + M + 6 = 47$$
$$2M + 15 = 47$$
$$2M = 32$$
$$M = 16$$

Thus, Mike is 16 now.

7. Let S = the number of student tickets sold. Then, $200 - S$ = the number of adult tickets sold.

Thus, the money from student tickets is $3S$, and the money received from adult tickets is $6(200 - S)$. Since a total of \$900 was collected,

$$3S + 6(200 - S) = 900$$
$$3S + 1,200 - 6S = 900$$
$$3S = 300$$
$$S = 100$$

Therefore, 100 student tickets were sold.

8. Let x = the amount invested at 12 percent. Then, since she invested a total of \$6,000, she must have invested \$6,000 $- x$ at 9 percent. And, since she received \$660 in interest, we have

$$12\%(x) + 9\%(6,000 - x) = 660 \quad \text{or,}$$
$$.12x + .09(6,000 - x) = 660$$
$$.12x + 540 - .09x = 660$$
$$.03x + 540 = 660$$
$$.03x = 120$$
$$x = \frac{120}{.03} = 4,000$$

She invested \$4,000 at 12 percent.

9. Let x = the number of days that it would take both pumps working together to fill the reservoir. In this time, the first pump will fill $\frac{x}{30}$ of the reservoir, and the second pump will fill $\frac{x}{20}$ of the reservoir. Thus, $\frac{1}{30} + \frac{1}{20} = \frac{1}{x}$. Multiply both sides by $60x$.

$$2x + 3x = 60$$
$$5x = 60$$
$$x = 12$$

It takes 12 days for both pumps to fill the reservoir together.

10. Let J = the time Jared needs to shovel the driveway alone.

    In 12 minutes, Brian can shovel $\frac{12}{21}$ of the driveway

    In 12 minutes, Peter can shovel $\frac{12}{84}$ of the driveway

    In 12 minutes, Jared can shovel $\frac{12}{J}$ of the driveway.

    Therefore,

    $$\frac{12}{21} + \frac{12}{84} + \frac{12}{J} = 1 \quad \text{Multiply both sides by } 84J.$$

    $$48J + 12J + 1{,}008 = 84J$$
    $$24J = 1{,}008$$
    $$J = 42$$

    Jared can shovel the driveway in 42 minutes.

11. Let B = Bobby's age. Then, $B + 3$ = Jimmy's age.

    In seven years, Bobby's age will be $B + 7$, and Jimmy's will be $B + 10$.

    Therefore, in 7 years, we will have

    $$(B + 7) + (B + 10) = 79$$
    $$2B + 17 = 79$$
    $$2B = 62$$
    $$B = 31$$

    Bobby is 31 now.

12. Let t = the amount of time each train travels. Then, the distance the freight train travels is $52t$, and the distance the passenger train travels is $84t$. Thus,

    $$52t + 84t = 680$$
    $$136t = 680$$
    $$t = 5$$

    The trains each travel for 5 hours, so they will be 680 miles apart at 5 p.m.

LITERAL EQUATIONS

An equation may have other letters in it besides the variable (or variables). Such an equation is called a *literal equation*. An illustration is $x + b = a$, with x being the variable. The solution of such an equation will not be a specific number but will involve letter symbols. Literal equations are solved by exactly the same methods as those involving numbers, but we must know which of the letters in the equation is to be considered the variable. Then the other letters are treated as constants.

Example 1

Solve $ax - 2bc = d$ for x.

$$ax = d + 2bc$$

$$x = \frac{d + 2bc}{a} \text{ if } a \neq 0$$

Example 2

Solve $ay - by = a^2 - b^2$ for y.

$y(a - b) = a^2 - b^2$	Factor out common term.
$y(a - b) = (a + b)(a - b)$	Factor expression on right side.
$y = a + b$	Divide each side by $a - b$ if $a \neq b$.

Example 3

Solve for S in the equation

$$\frac{1}{R} = \frac{1}{S} + \frac{1}{T}$$

Multiply every term by RST, the LCD:

$$ST = RT + RS$$
$$ST - RS = RT$$
$$S(T - R) = RT$$
$$S = \frac{RT}{T - R} \text{ if } T \neq R$$

Quadratic Equations

An equation containing the square of an unknown quantity is called a *quadratic* equation. One way of solving such an equation is by factoring. If the product of two expressions is zero, at least one of the expressions must be zero.

Example 1

Solve $y^2 + 2y = 0$.

$y(y + 2) = 0$ Remove common factor.

$y = 0$ or $y + 2 = 0$ Since product is 0, at least one of the factors must be 0.

$y = 0$ or $y = -2$

Check by substituting both values in the original equation:

$$(0)^2 + 2(0) = 0$$
$$(-2)^2 + 2(-2) = 4 - 4 = 0$$

In this case there are two solutions.

Example 2

Solve $x^2 + 7x + 10 = 0$.

$x^2 + 7x + 10 = (x + 5)(x + 2) = 0$
$x + 5 = 0$ or $x + 2 = 0$
$x = -5$ or $x = -2$

Check:

$(-5)^2 + 7(-5) + 10 = 25 - 35 + 10 = 0$
$(-2)^2 + 7(-2) + 10 = 4 - 14 + 10 = 0$

Not all quadratic equations can be factored using only integers, but solutions can usually be found by means of a formula. A quadratic equation may have two solutions, one solution, or occasionally no real solutions. If the quadratic equation is in the form $Ax^2 + Bx + C = 0$, x can be found from the following formula:

$$x = \frac{-B \pm \sqrt{B^2 - 4AC}}{2A}$$

Example 3

Solve $2x^2 + 5x + 2 = 0$ by formula. Assume $A = 2$, $B = 5$, and $C = 2$.

$$x = \frac{-5 \pm \sqrt{5^2 - 4(2)(2)}}{2(2)}$$

$$= \frac{-5 \pm \sqrt{25 - 16}}{4}$$

$$= \frac{-5 \pm \sqrt{9}}{4}$$

$$= \frac{-5 \pm 3}{4}$$

This yields two solutions:

$$x = \frac{-5 + 3}{4} = \frac{-2}{4} = -\frac{1}{2} \quad \text{and}$$

$$x = \frac{-5 - 3}{4} = \frac{-8}{4} = -2$$

So far, each quadratic we have solved has had two distinct answers, but an equation may have a single answer (repeated), as in

$x^2 + 4x + 4 = 0$

$(x + 2)(x + 2) = 0$

$x + 2 = 0$ and $x + 2 = 0$

$x = -2$ and $x = -2$

The only solution is -2.

It is also possible for a quadratic equation to have no real solution at all.

Example 4

If we attempt to solve $x^2 + x + 1 = 0$ by formula, we get:

$$x = \frac{-1 \pm \sqrt{1 - 4(1)(1)}}{2} = \frac{-1 \pm \sqrt{-3}}{2}$$ Since $\sqrt{-3}$ is not defined, this quadratic has no real answer.

Rewriting Equations

Certain equations written with a variable in the denominator can be rewritten as quadratics.

Example

Solve $-\dfrac{4}{x} + 5 = x$

$-4 + 5x = x^2$	Multiply both sides by x.
$-x^2 + 5x - 4 = 0$	Collect terms on one side of equals and set sum equal to 0.
$x^2 - 5x + 4 = 0$	Multiply both sides by -1.
$(x - 4)(x - 1) = 0$	Factor
$x - 4 = 0$ or $x - 1 = 0$	
$x = 4$ or $x = 1$	

Check the result by substitution:

$$-\dfrac{4}{4} + 5 \stackrel{?}{=} 4 \quad \text{and} \quad -\dfrac{4}{1} + 5 \stackrel{?}{=} 1$$

$$-1 + 5 = 4 \qquad\qquad -4 + 5 = 1$$

Some equations containing a radical sign can also be converted into a quadratic equation. The solution of this type of problem depends on the principle that

If $A = B$ then $A^2 = B^2$
and If $A^2 = B^2$ then $A = B$ or $A = -B$

Equations Involving Square Roots

To solve equations in which the variable appears under a square root sign, begin by manipulating the equation so that the square root is alone on one side. Then square both sides of the equation. Since squares and square roots are inverses, the square root will be eliminated from the equation.

Example 1

Solve $\sqrt{12x + 4} + 2 = 6$

Rewrite the equation as $\sqrt{12x + 4} = 4$. Now square both sides.

$$(\sqrt{12x + 4})^2 = 4^2$$
$$12x + 4 = 16$$
$$12x = 12$$
$$x = 1$$

It is easy to check that 1 is a solution to the equation by plugging the 1 into the original equation. However, sometimes when we use this procedure, the solution obtained will not solve the original equation. Thus, it is crucial to check your answer to all square root equations.

Example 2

Solve $y = \sqrt{3y + 4}$.

$$y = \sqrt{3y + 4}$$
$$y^2 = 3y + 4$$
$$y^2 - 3y - 4 = 0$$
$$(y - 4)(y + 1) = 0$$
$$y = 4 \text{ or } y = -1$$

Check by substituting values into the original equation:

$$4 \stackrel{?}{=} \sqrt{3(4) + 4} \text{ and}$$
$$-1 \stackrel{?}{=} \sqrt{3(-1) + 4}$$
$$4 \stackrel{?}{=} \sqrt{16} \quad -1 \stackrel{?}{=} \sqrt{1}$$
$$4 = 4 \quad\quad -1 \neq 1$$

The single solution is $y = 4$; the false root $y = -1$ was introduced when the original equation was squared.

QUIZ

EQUATION-SOLVING PROBLEMS

Solve the following equations for the variable indicated.

1. Solve for c: $A = \frac{1}{2}b(b + c)$
2. Solve for b_2: $2A = (b_1 + b_2)b$
3. Solve for w: $aw - b = cw + d$
4. Solve for d: $\left(\frac{a}{b}\right) = \left(\frac{c}{d}\right)$
5. Solve for x: $10x^2 = 5x$
6. Solve for x: $2x^2 - x = 21$
7. Solve for x: $3x^2 - 12 = x(1 + 2x)$
8. Solve for x: $2\sqrt{x + 5} = 8$
9. Solve for x: $5x^2 = 36 + x^2$
10. Solve for x: $4\sqrt{\frac{2x}{3}} = 48$
11. Solve for x: $3x^2 - x - 4 = 0$
12. Solve $\frac{q}{x} + \frac{p}{x} = 1$ for x
13. Solve for x: $3x^2 - 5 = 0$

ALGEBRA

SOLUTIONS

1. $$A = \frac{1}{2}h(b + c)$$
    $$2A = hb + hc$$
    $$2A - hb = hc$$
    $$c = \left(\frac{2A - hb}{h}\right)$$

2. $$2A = (b_1 + b_2)h$$
    $$2A = b_1h + b_2h$$
    $$2A - b_1h = b_2h$$
    $$\frac{(2A - b_1h)}{h} = b_2$$

3. $$aw - b = cw + d$$
    $$aw - cw = b + d$$
    $$w(a - c) = b + d$$
    $$w = \frac{(b + d)}{(a - c)}$$

4. $\left(\frac{a}{b}\right) = \left(\frac{c}{d}\right)$ Cross multiply
    $$ad = bc$$
    $$d = \frac{(bc)}{a}$$

5. $$10x^2 = 5x$$
    $$10x^2 - 5x = 0$$
    $$5x(2x - 1) = 0$$
    $$x = 0, \frac{1}{2}$$

6. $$2x^2 - x = 21$$
    $$2x^2 - x - 21 = 0$$
    $$(2x - 7)(x + 3) = 0$$
    $$x = -3, \frac{7}{2}$$

7. $$3x^2 - 12 = x(1 + 2x)$$
    $$3x^2 - 12 = x + 2x^2$$
    $$x^2 - x - 12 = 0$$
    $(x - 4)(x + 3) = 0$ Thus, $x = 4$ or -3.

8. $2\sqrt{x+5} = 8$
    $\sqrt{x+5} = 4$
    $(\sqrt{x+5})^2 = 4^2$
    $x + 5 = 16$
    $x = 11$

9. $5x^2 = 36 + x^2$
    $4x^2 - 36 = 0$
    $4(x^2 - 9) = 0$
    $4(x + 3)(x - 3) = 0$ Thus, $x = -3$ or $+3$.

10. Begin by dividing both sides by 4 to get $\sqrt{\frac{2x}{3}} = 12$. Then, square both sides:

    $\left(\sqrt{\frac{2x}{3}}\right)^2 = 12^2$

    $\frac{2x}{3} = 144$ Now, multiply both sides by 3.

    $2x = 432$
    $x = 216$

11. $3x^2 - x - 4 = 0$ Here, A = 3, B = -1, and C = -4. Using the quadratic formula, we get:

    $x = \frac{-B \pm \sqrt{B^2 - 4AC}}{2A} = \frac{1 \pm \sqrt{1 - 4(3)(-4)}}{6}$
    $= \frac{1 \pm \sqrt{1 + 48}}{6} = \frac{1 \pm \sqrt{49}}{6} = \frac{1 \pm 7}{6} = \frac{8}{6}, \frac{-6}{6}$

    Thus, $w = \frac{4}{3}$ or -1. Note that this equation could have been solved as well by factoring. The quadratic formula, however, can be used to solve all quadratic equations, including those that cannot be factored.

12. $\frac{q}{x} + \frac{p}{x} = 1$ Multiply both sides by x to clear the fraction, and obtain $q + p = x$.

13. $3x^2 - 5 = 0$

    This equation can easily be solved for x by first solving for x^2 and then taking the square root of both sides.

    $3x^2 = 5$

    $x^2 = \frac{5}{3}$

    $\sqrt{x^2} = \pm\sqrt{\frac{5}{3}}$ Since $\sqrt{x^2} = x$, we have $x = \pm\sqrt{\frac{5}{3}}$

LINEAR INEQUALITIES

For each of the sets of numbers we have considered, we have established an ordering of the members of the set by defining what it means to say that one number is greater than the other. Every number we have considered can be represented by a point on a number line.

An *algebraic inequality* is a statement that one algebraic expression is greater than (or less than) another algebraic expression. If all the variables in the inequality are raised to the first power, the inequality is said to be a *linear inequality*. We solve the inequality by reducing it to a simpler inequality whose solution is apparent. The answer is not unique, as it is in an equation, since a great number of values may satisfy the inequality.

There are three rules for producing equivalent inequalities:

1. The same quantity can be added or subtracted from each side of an inequality.

2. Each side of an inequality can be multiplied or divided by the same *positive* quantity.

3. If each side of an inequality is multiplied or divided by the same *negative* quantity, the sign of the inequality must be reversed so that the new inequality is equivalent to the first.

Example 1

Solve $5x - 5 > -9 + 3x$.

$5x > -4 + 3x$ Add 5 to each side.
$2x > -4$ Subtract $3x$ from each side.
$x > -2$ Divide by $+2$.

Any number greater than -2 is a solution to this inequality.

Example 2

Solve $2x - 12 < 5x - 3$.

$2x < 5x + 9$ Add 12 to each side.
$-3x < 9$ Subtract $5x$ from each side.
$x > -3$ Divide each side by -3, changing sign of inequality.

Any number greater than -3 (for example, $-2\frac{1}{2}$, 0, 1, or 4) is a solution to this inequality.

Example 3

$$\frac{x}{3} - \frac{x}{2} > 1$$

Begin by multiplying both sides by 6 to clear the fractions. We then obtain

$2x - 3x > 6$
$-x > 6$

Now, divide both sides by -1, and reverse the inequality.

$x < -6$

Linear Equations in Two Unknowns

Graphing Equations

The number line is useful in picturing the values of one variable. When two variables are involved, a coordinate system is effective. The Cartesian coordinate system is constructed by placing a vertical number line and a horizontal number line on a plane so that the lines intersect at their zero points. This meeting place is called the *origin*. The horizontal number line is called the *x* axis, and the vertical number line (with positive numbers above the *x* axis) is called the *y* axis. Points in the plane correspond to ordered pairs of real numbers.

Example 1

The points in this example are:

x	y
0	0
1	1
3	−1
−2	−2
−2	1

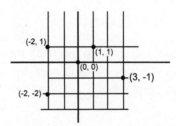

A first-degree equation in two variables is an equation that can be written in the form $ax + by = c$, where a, b, and c are constants. *First-degree* means that x and y appear to the first power. *Linear* refers to the graph of the solutions (x, y) of the equation, which is a straight line. We have already discussed linear equations of one variable.

Example 2

Graph the line $y = 2x - 4$.

First make a table and select small integral values of x. Find the value of each corresponding y and write it in the table:

x	y
0	-4
1	-2
2	0
3	2

If $x = 1$, for example, $y = 2(1) - 4 = -2$. Then plot the four points on a coordinate system. It is not necessary to have four points; two would do, since two points determine a line, but plotting three or more points reduces the possibility of error.

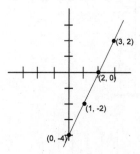

After the points have been plotted (placed on the graph), draw a line through the points and extend it in both directions. This line represents the equation $y = 2x - 4$.

Solving Simultaneous Linear Equations

Two linear equations can be solved together (simultaneously) to yield an answer (x, y) if it exists. On the coordinate system, this amounts to drawing the graphs of two lines and finding their point of intersection. If the lines are parallel and therefore never meet, no solution exists.

Simultaneous linear equations can be solved in the following manner without drawing graphs. From the first equation, find the value of one variable in terms of the other; substitute this value into the second equation. The second equation is now a linear equation in one variable and can be solved. After the numerical value of the one variable has been found, substitute that value into the first equation to find the value of the second variable. Check the results by putting both values into the second equation.

ALGEBRA

Example 1

Solve the system

$2x + y = 3$
$4x - y = 0$

From the first equation, $y = 3 - 2x$. Substitute this value of y into the second equation to get:

$4x - (3 - 2x) = 0$
$4x - 3 + 2x = 0$
$6x = 3$
$x = \frac{1}{2}$

Substitute $x = \frac{1}{2}$ into the first of the original equations:

$2\left(\frac{1}{2}\right) + y = 3$
$1 + y = 3$
$y = 2$

Check by substituting both x and y values into the second equation:

$4\left(\frac{1}{2}\right) - 2 = 0$
$2 - 2 = 0$

Example 2

The sum of two numbers is 87 and their difference is 13. What are the numbers?

Let $x =$ the larger of the two numbers and y the smaller. Then,

$x + y = 87$
$x - y = 13.$

Rewrite the second equation as $x = y + 13$ and plug it into the first equation.

$(y + 13) + y = 87$
$2y + 13 = 87$
$2y = 74$
$y = 37$

Then, $x = 13 + 37 = 50$.

The numbers are 50 and 37.

Example 3

A change-making machine contains $30 in dimes and quarters. There are 150 coins in the machine. Find the number of each type of coin.

Let x = number of dimes and y = number of quarters. Then:

$x + y = 150$

Since $.25y$ is the product of a quarter of a dollar and the number of quarters and $.10x$ is the amount of money in dimes,

$.10x + .25y = 30$

Multiply the last equation by 100 to eliminate the decimal points:

$10x + 25y = 3,000$

From the first equation, $y = 150 - x$. Substitute this value into the equivalent form of the second equation.

$10x + 25(150 - x) = 3,000$
$-15x = -750$
$x = 50$

This is the number of dimes. Substitute this value into $x + y = 150$ to find the number of quarters, $y = 100$.

Check:

$.10(50) + .25(100) = 30$
$\$5 + \$25 = \$30$

Exponential Equations

On your test, you may also have to solve simple exponential equations. An exponential equation is an equation whose variable appears in a exponent. Such equations can be solved by algebraic means if it is possible to express both sides of the equation as powers of the same base.

Example 1

Solve $5^{2x-1} = 25$.

Rewrite the equation as $5^{2x-1} = 5^2$. Then it must be true that $2x - 1 = 2$. This means that $x = \frac{3}{2}$.

Example 2

Solve $9^{x+3} = 27^{2x}$.

Rewrite the left side of the equation as $(3^2)^{x+3} = 3^{2x+6}$. Rewrite the right side of the equation as $(3^3)^{2x} = 3^{6x}$. Then, it must be true that $2x + 6 = 6x$. This means that $x = \frac{3}{2}$.

Exponential equations, in which the bases cannot both be changed to the same number, can be solved by using logarithms.

QUIZ

LINEAR INEQUALITIES AND EQUATIONS PROBLEMS

1. Solve for x: $12 - 2x > 4$

2. Solve for x: $\left(\dfrac{x}{6}\right) - \left(\dfrac{x}{2}\right) < 1$

3. Solve for x: $108x < 15(6x + 12)$

4. Solve for a: $4a - 9 > 9a - 24$

5. Solve for z: $6z + 1 \leq 3(z - 2)$

6. Find the common solution:
   $y = 3x + 1$
   $x + y = 9$

7. Find the common solution:
   $2x + y = 8$
   $x - y = 1$

8. Find the common solution:
   $3x + 2y = 11$
   $5x - 4y = 11$

9. Solve for a common solution:
   $5x + 3y = 28$
   $7x - 2y = 2$

10. The sum of two numbers is 45 and their difference is 11. What are the two numbers?

11. Three binders and four notebooks cost a total of $4.32. One binder and five notebooks cost $3.97. What is the cost of one notebook?

12. A printer and monitor together cost $356. The monitor costs $20 more than two times the printer. How much do the printer and monitor cost separately?

ALGEBRA

SOLUTIONS

1. $12 - 2x > 4$
    $-2x > -8$ Divide by -2, and flip inequality sign.
    $x < 4$

2. $\left(\dfrac{x}{6}\right) - \left(\dfrac{x}{2}\right) < 1$ Multiply both sides by 6.
    $x - 3x < 6$
    $-2x < 6$ Divide by -2, and flip the inequality sign.
    $x > -3$

3. $108x < 15(6x + 12)$
    $108x < 90x + 180$
    $18x < 180$
    $x < 10$

4. $4a - 9 > 9a - 24$
    $-5a > -15$ Divide by -5, and reverse the inequality sign.
    $a < 3$

5. $6z + 1 \leq 3(z - 2)$
    $6z + 1 \leq 3z - 6$
    $3z \leq -7$
    $z \leq \dfrac{-7}{3}$

    Note that even though the answer is negative, we do not reverse the inequality sign since we never multiplied or divided by a negative number.

6. $y = 3x + 1$
    $x + y = 9$
    Begin by substituting $y = 3x + 1$ into the second equation.

    $x + (3x + 1) = 9$
    $4x + 1 = 9$
    $4x = 8$
    $x = 2$
    If $x = 2$, $y = 3(2) + 1 = 6 + 1 = 7$.

7. $2x + y = 8$
   $x - y = 1$

   From the second equation, we can see $x = y + 1$. Then, substituting into the first equation:

   $2(y + 1) + y = 8$
   $3y + 2 = 8$
   $3y = 6$
   $y = 2$

   If $y = 2$, then $x = y + 1 = 2 + 1 = 3$.

8. $3x + 2y = 11$
   $5x - 4y = 11$

   Multiply the top equation by 2.
   $2(3x + 2y) = 2(11)$
   $5x - 4y = 11$

   $6x + 4y = 22$
   $\underline{5x - 4y = 11}$
   $11x = 33$
   $x = 3$

   Now, substitute this value for x in the first equation.
   $3(3) + 2y = 11$
   $9 + 2y = 11$
   $2y = 2$
   $y = 1$

   The common solution is (3, 1).

9. $5x + 3y = 28$
   $7x - 2y = 2$

   Multiply the first equation by 2, and the second equation by 3.

   $2(5x + 3y) = 2(28)$
   $3(7x - 2y) = 3(2)$ Thus,

   $10x + 6y = 56$
   $\underline{21x - 6y = 6}$ Add the equations.
   $31x = 62$
   $x = 2$

   Now, solve for y by plugging $x = 2$ into (say) the second equation.

   $7(2) - 2y = 2$
   $14 - 2y = 2$
   $-2y = -12$
   $y = 6$

   Thus, the common solution is (2, 6).

10. Let x = the larger of the two numbers.
    Let y = the smaller of the two numbers.
    Then, we have:

    $x + y = 45$
    $x - y = 11$

    Add the two equations.

    $x + y = 45$
    $\underline{x - y = 11}$
    $2x = 56$
    $x = 28$

    If x is 28 and the numbers differ by 11, then $y = 17$.

    The numbers are 28 and 17.

11. Let B = the cost of a binder.
    Let N = the cost of a notebook.

    Then, we have:

    $3B + 4N = 4.32$
    $1B + 5N = 3.97$

    Multiply the second equation by -3:

    $3B + 4N = 4.32$
    $-3(1B + 5N) = (3.97)(-3)$ or,

    $3B + 4N = 4.32$
    $\underline{-3B - 15N = -11.91}$
    $-11N = -7.59$
    $N = 0.69$

    Thus, the cost of a notebook is $ 0.69.

12. A printer and monitor together cost $356. The monitor costs $20 more than two times the printer.

    Let P = the cost of the printer.
    Let M = the cost of the monitor. Then,

    $P + M = 356$
    $M = 20 + 2P$

    Substituting for M in the first equation, we get:
    $P + (20 + 2P) = 356$
    $3P + 20 = 356$
    $3P = 336$
    $P = 112$

    Then, $M = 20 + 2(112) = 244$.

    The printer costs $112, and the monitor costs $244.

Ratio and Proportion

Many problems in arithmetic and algebra can be solved using the concept of *ratio* to compare numbers. The ratio of a to b is the fraction $\frac{a}{b}$. If the two ratios $\frac{a}{b}$ and $\frac{c}{d}$ represent the same comparison, we write:

$$\frac{a}{b} = \frac{c}{d}$$

This equation (statement of equality) is called a *proportion*. A proportion states the equivalence of two different expressions for the same ratio.

Example 1

In a class of 39 students, 17 are men. Find the ratio of men to women.

39 students − 17 men = 22 women

Ratio of men to women is 17/22, also written 17:22.

Example 2

The scale on a map is $\frac{3}{4}$ inch = 12 miles. If the distance between City A and City B on the map is $4\frac{1}{2}$ inches, how far apart are the two cities actually?

Let $x =$ the distance between the two cities in miles.

Begin by writing a proportion that compares inches to miles.

$$\frac{\text{Inches} \rightarrow}{\text{Miles} \rightarrow} \frac{\frac{3}{4}}{12} = \frac{\frac{9}{2}}{x} \quad \text{Cross multiply to solve the equation.}$$

$$\left(\frac{3}{4}\right)x = 12\left(\frac{9}{2}\right)$$

$$\left(\frac{3}{4}\right)x = 54 \quad \text{Multiply by 4}$$

$$3x = 216$$

$$x = 72$$

The two cities are 72 miles apart.

Example 3

A fertilizer contains 3 parts nitrogen, 2 parts potash, and 2 parts phosphate by weight. How many pounds of fertilizer will contain 60 pounds of nitrogen?

The ratio of pounds of nitrogen to pounds of fertilizer is 3 to $3 + 2 + 2 = \frac{3}{7}$. Let x be the number of pounds of mixture. Then:

$$\frac{3}{7} = \frac{60}{x}$$

Multiply both sides of the equation by $7x$ to get:

$3x = 420$

$x = 140$ pounds

Variation

The concept of variation is useful for describing a number of situations that arise in science. If x and y are variables, then *y is said to vary directly as x* if there is a nonzero constant k such that $y = kx$.

Similarly, we say that *y varies inversely as x* if $y = \frac{k}{x}$ for some nonzero constant k. To say that *y varies inversely as the square of x* means that $y = \frac{k}{x^2}$ for some nonzero constant k. Finally, to say *y varies jointly as s and t* means that $y = kst$ for some nonzero constant k.

Example

Boyle's law says that, for an enclosed gas at a constant temperature, the pressure p varies inversely as the volume v. If $v = 10$ cubic inches when $p = 8$ pounds per square inch, find v when $p = 12$ pounds per square inch.

Since p varies inversely as v, we have $p = \frac{k}{v}$, for some value of k. We know that when $p = 8$, $v = 10$, so $8 = \frac{k}{10}$. This tells us that the value of k is 80, and we have $p = \frac{80}{v}$. For $p = 12$, we have $12 = \frac{80}{v}$, or $v = \frac{80}{12} = 6.67$ cubic inches.

COMPUTING AVERAGES AND MEDIANS

Mean

Several statistical measures are used frequently. One of them is the *average* or *arithmetic mean*. To find the average of N numbers, add the numbers and divide the sum by N.

Example 1

Seven students attained test scores of 62, 80, 60, 30, 50, 90, and 20. What was the average test score for the group?

$62 + 80 + 60 + 30 + 50 + 90 + 20 = 392$

Since there are 7 scores, the average score is:

$$\frac{392}{7} = 56$$

Example 2

Brian has scores of 88, 87, and 92 on his first three tests. What grade must he get on his next test to have an overall average of 90?

Let $x =$ the grade that he needs to get. Then we have:

$$\frac{88 + 87 + 92 + x}{4} = 90 \quad \text{Multiply by 4 to clear the fraction.}$$

$88 + 87 + 92 + x = 360$

$267 + x = 360$

$x = 93$

Brian needs to get a 93 on his next test.

Example 3

Joan allotted herself a budget of $50 a week, on the average, for expenses. One week she spent $35, the next $60, and the third $40. How much can she spend in the fourth week without exceeding her budget?

Let x be the amount spent in the fourth week. Then:

$$\frac{35 + 60 + 40 + x}{4} = 50$$
$$35 + 60 + 40 + x = 200$$
$$135 + x = 200$$
$$x = 65$$

She can spend $65 in the fourth week.

Median

If a set of numbers is arranged in order, the number in the middle is called the *median*.

Example

Find the median test score of 62, 80, 60, 30, 50, 90, and 20. Arrange the numbers in increasing (or decreasing) order.

20, 30, 50, 60, 62, 80, 90

Since 60 is the number in the middle, it is the median. It is not the same as the arithmetic mean, which is 56.

If the number of scores is an even number, the median is the arithmetic mean of the middle two scores.

PLANE GEOMETRY

There are a significant number of plane geometry questions on the Level IC Math test. While plane geometry is not specifically tested on the Level IIC test, plane geometry is crucial background for a number of other topics, such as coordinate geometry and trigonometry, that *are* on the Level IIC test. Thus, it is recommended that all readers review the following plane geometry sections.

Plane geometry is the science of measurement. Certain assumptions are made about undefined quantities called points, lines, and planes, and then logical deductions about relationships between figures composed of lines, angles, and portions of planes are made based on these assumptions. The process of making the logical deductions is called a *proof*. In this summary, we are not making any proofs but are giving the definitions frequently used in geometry and stating relationships that are the results of proofs.

LINES AND ANGLES

Angles

A line in geometry is always a straight line. When two straight lines meet at a point, they form an *angle*. The lines are called *sides* or *rays* of the angle, and the point is called the *vertex*. The symbol for angle is ∠. When no other angle shares the same vertex, the name of the angle is the name given to the vertex, as in angle *A*:

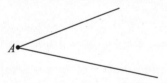

An angle may be named with three letters. In the following example, *B* is a point on one side and *C* is a point on the other. In this case, the name of the vertex must be the middle letter, and we have angle *BAC*.

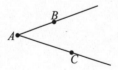

Occasionally, an angle is named by a number or small letter placed in the angle.

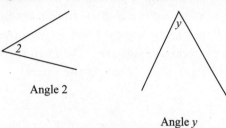

Angle 2

Angle y

Angles are usually measured in degrees. An angle of 30 degrees, written 30°, is an angle whose measure is 30 degrees. Degrees are divided into minutes; 60' (read "minutes") = 1°. Minutes are further divided into seconds; 60" (read "seconds") = 1'.

Vertical Angles

When two lines intersect, four angles are formed. The angles opposite each other are called *vertical angles* and are equal to each other.

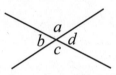

 a and c are vertical angles.
 ∠a = ∠c.
 b and d are vertical angles.
 ∠b = ∠d.

Straight Angle

A *straight angle* has its sides lying along a straight line. It is always equal to 180°.

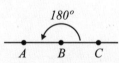

∠ABC = ∠B = 180°.
∠B is a straight angle.

Adjacent Angles

Two angles are *adjacent* if they share the same vertex and a common side but no angle is inside another angle. ∠ABC and ∠CBD are adjacent angles. Even though they share a common vertex B and a common side AB, ∠ABD and ∠ABC are not adjacent angles because one angle is inside the other.

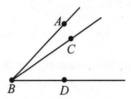

Supplementary Angles

If the sum of two angles is a straight angle (180°), the two angles are *supplementary* and each angle is the supplement of the other.

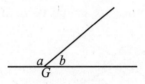

∠G is a straight angle = 180°.
∠a + ∠b = 180°
∠a and ∠b are supplementary angles.

Right Angles

If two supplementary angles are equal, they are both *right angles*. A right angle is one half a straight angle. Its measure is 90°. A right angle is symbolized by ∟.

∠G is a straight angle.
∠b + ∠a = ∠G, and ∠a = ∠b. ∠a and ∠b are right angles.

Complementary Angles

Complementary angles are two angles whose sum is a right angle (90°).

∠Y is a right angle.
∠a + ∠b = ∠Y = 90°.
∠a and ∠b are complementary angles.

Acute Angles

Acute angles are angles whose measure is less than 90°. No two acute angles can be supplementary angles. Two acute angles can be complementary angles.

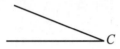

∠C is an acute angle.

Obtuse Angles

Obtuse angles are angles that are greater than 90° and less than 180°.

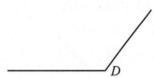

∠D is an obtuse angle.

Example 1

In the figure, what is the value of x?

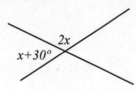

Since the two labeled angles are supplementary angles, their sum is 180°.

$$(x + 30°) + 2x = 180°$$
$$3x = 150°$$
$$x = 50°$$

Example 2

Find the value of x in the figure.

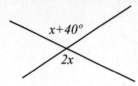

Since the two labeled angles are vertical angles, they are equal.

$x + 40° = 2x$
$40° = x$

Example 3

If angle Y is a right angle and angle b measures $30°15'$, what does angle a measure?

Since angle Y is a right angle, angles a and b are complementary angles and their sum is $90°$.

$\angle a + \angle b = 90°$
$\angle a + 30°15' = 90°$
$\angle a = 59°45'$

Example 4

In the figure below, what is the value of *x*?

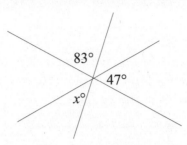

The angle that is vertical to the angle labeled *x*° also has a measure of *x*°. This angle, along with those labeled 83° and 47°, form a straight line and are thus supplementary. Therefore:

$$83 + 47 + x = 180$$
$$130 + x = 180$$
$$x = 50$$

The value of *x* is 50°.

LINES

A *line* in geometry is always assumed to be a straight line. It extends infinitely far in both directions. It can be determined if two of its points are known. It can be expressed in terms of the two points, which are written as capital letters. The following line is called *AB*.

Or a line may be given one name with a small letter. The following line is called line *k*.

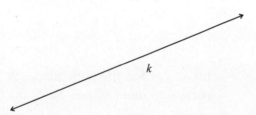

A *line segment* is a part of a line between two *endpoints*. It is named by its endpoints, for example, *A* and *B*.

AB is a line segment.
It has a definite length.

If point *P* is on the line and is the same distance from point *A* as from point *B*, then *P* is the *midpoint* of segment *AB*. When we say *AP = PB*, we mean that the two line segments have the same length.

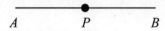

A part of a line with one endpoint is called a *ray*. AC is a ray, of which *A* is an endpoint. The ray extends infinitely in the direction away from the endpoint.

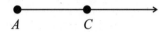

Parallel Lines

Two lines meet or intersect if there is one point that is on both lines. Two different lines may either intersect at one point or never meet, but they can never meet at more than one point.

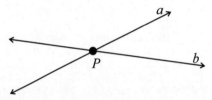

Two lines in the same plane that never meet no matter how far they are extended are said to be *parallel,* for which the symbol is ∥. In the following diagram, *a* ∥ *b*.

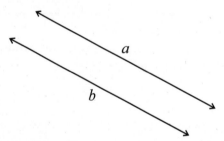

If two lines in the same plane are parallel to a third line, they are parallel to one another. Since *a* ∥ *b* and *b* ∥ *c*, we know that *a* ∥ *c*.

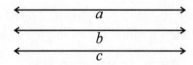

Two lines that meet one another at right angles are said to be *perpendicular,* for which the symbol is ⊥. Line *a* is perpendicular to line *b*.

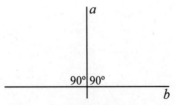

Two lines in the same plane that are perpendicular to the same line are parallel to each other.

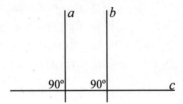

Line *a* ⊥ line *c* and line *b* ⊥ line *c*.
Therefore, *a* ∥ *b*.

A line intersecting two other lines is called a *transversal*. Line *c* is a transversal intersecting lines *a* and *b*.

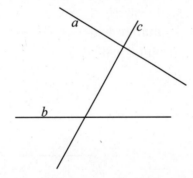

The transversal and the two given lines form eight angles. The four angles between the given lines are called *interior angles;* the four angles outside the given lines are called *exterior angles.* If two angles are on opposite sides of the transversal, they are called *alternate angles.*

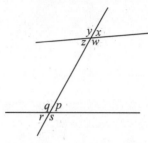

∠z, ∠w, ∠q, and ∠p are interior angles.
∠y, ∠x, ∠s, and ∠r are exterior angles.
∠z and ∠p are alternate interior angles; so are ∠w and ∠q.
∠y and ∠s are alternate exterior angles; so are ∠x and ∠r.

Pairs of *corresponding* angles are ∠y and ∠q, ∠z and ∠r, ∠x and ∠p, and, ∠w and ∠s. Corresponding angles are sometimes called exterior-interior angles.

When the two given lines cut by a transversal are parallel lines:

1. the corresponding angles are equal.
2. the alternate interior angles are equal.
3. the alternate exterior angles are equal.
4. interior angles on the same side of the transversal are supplementary.

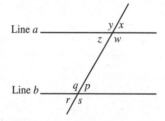

If line *a* is parallel to line *b:*

1. ∠y = ∠q, ∠z = ∠r, ∠x = ∠p, and ∠w = ∠s.
2. ∠z = ∠p and ∠w = ∠q.
3. ∠y = ∠s and ∠x = ∠r.
4. ∠z + ∠q = 180° and ∠p + ∠w = 180°.

Because vertical angles are equal, ∠p = ∠r, ∠q = ∠s, ∠y = ∠w, and ∠x = ∠z. If any one of the four conditions for equality of angles holds true, the lines are parallel; that is, if two lines are cut by a transversal and one pair of the corresponding angles is equal, the lines are parallel. If a pair of alternate interior angles or a pair of alternate exterior angles is equal, the lines are parallel. If interior angles on the same side of the transversal are supplementary, the lines are parallel.

Example 1

In the figure below, two parallel lines are cut by a transversal. Find the measure of angle y.

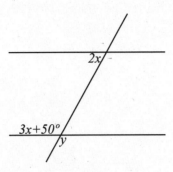

The two labeled angles are supplementary.

$$2x + (3x + 50°) = 180°$$
$$5x = 130°$$
$$x = 26°$$

Since $\angle y$ is vertical to the angle whose measure is $3x + 50°$, it has the same measure.

$$y = 3x + 50° = 3(26°) + 50° = 128°$$

Example 2

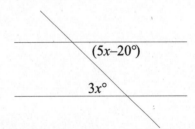

In the figure above, two parallel lines are cut by a transversal. Find the measure of angle x.

The two labeled angles are alternate interior angles and thus are congruent. Therefore:

$$5x - 20 = 3x$$
$$2x = 20$$
$$x = 10$$

The measure of angle x is 10°.

Polygons

A *polygon* is a closed plane figure composed of line segments joined together at points called *vertices* (singular, *vertex*). A polygon is usually named by giving its vertices in order.

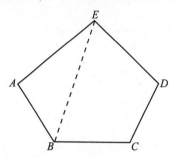

Polygon *ABCDE*

In the figure, points *A, B, C, D,* and *E* are the vertices, and the sides are *AB, BC, CD, DE,* and *EA*. *AB* and *BC* are *adjacent* sides, and *A* and *B* are adjacent vertices. A *diagonal* of a polygon is a line segment joining any two nonadjacent vertices. *EB* is a diagonal.

Polygons are named according to the number of sides or angles. A *triangle* is a polygon with three sides, a *quadrilateral* a polygon with four sides, a *pentagon* a polygon with five sides, and a *hexagon* a polygon with six sides. The number of sides is always equal to the number of angles.

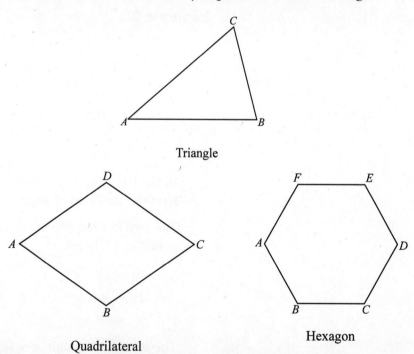

The *perimeter* of a polygon is the sum of the lengths of its sides. If the polygon is regular (all sides equal and all angles equal), the perimeter is the product of the length of *one* side and the number of sides.

Congruent and Similar Polygons

If two polygons have equal corresponding angles and equal corresponding sides, they are said to be *congruent.* Congruent polygons have the same size and shape. They are the same in all respects except possibly position. The symbol for congruence is ≅.

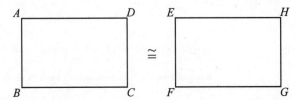

When two sides of congruent or different polygons are equal, we indicate the fact by drawing the same number of short lines through the equal sides.

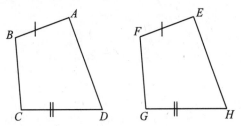

This indicates that *AB = EF* and *CD = GH*.

Two polygons with equal corresponding angles and corresponding sides in proportion are said to be *similar.* The symbol for similar is ∼.

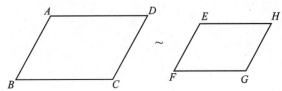

Similar figures have the same shape but not necessarily the same size.

A *regular polygon* is a polygon whose sides are equal and whose angles are equal.

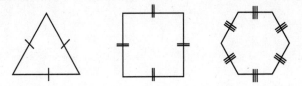

TRIANGLES

A *triangle* is a polygon of three sides. Triangles are classified by measuring their sides and angles. The sum of the angles of a plane triangle is always 180°. The symbol for a triangle is Δ. The sum of any two sides of a triangle is always greater than the third side.

Equilateral

Equilateral triangles have equal sides and equal angles. Each angle measures 60° because $\frac{1}{3}(180°) = 60°$.

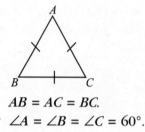

AB = AC = BC.
∠A = ∠B = ∠C = 60°.

Isosceles

Isosceles triangles have two equal sides. The angles opposite the equal sides are equal. The two equal angles are sometimes called the *base* angles, and the third angle is called the *vertex* angle. Note that an equilateral triangle is isosceles.

FG = FH.
FG ≠ GH.
∠G = ∠H.
∠F is the vertex angle.
∠G and ∠H are base angles.

Scalene

Scalene triangles have all three sides of different length and all angles of different measure. In scalene triangles, the shortest side is opposite the angle of smallest measure, and the longest side is opposite the angle of greatest measure.

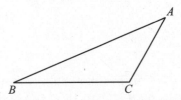

$AB > BC > CA$; therefore, $\angle C > \angle A > \angle B$.

Example 1

In triangle XYZ, $\angle Y$ is twice $\angle X$, and $\angle Z$ is 40° more than $\angle Y$. How many degrees are in the three angles?

Solve this problem just as you would an algebraic word problem, remembering that there are 180° in a triangle.

Let x = the number of degrees in $\angle X$.

Then $2x$ = the number of degrees in $\angle Y$

and $2x + 40$ = the number of degrees in $\angle Z$.

Thus,

$$x + 2x + (2x + 40) = 180$$
$$5x + 40 = 180$$
$$5x = 140$$
$$x = 28°$$

Therefore, the measure of $\angle X$ is 28°, the measure of $\angle Y$ is 56°, and the measure of $\angle Z$ is 96°.

Example 2

In the figure below, the two lines are parallel. What is the value of *x*?

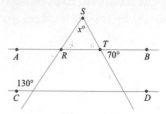

Corresponding angles are equal, so ∠*ARS* is also 130°. ∠*SRT* is the supplement of ∠*ARS* and thus is 50°. By the property of vertical angles, we have ∠*STR* = 70°. Finally, since the sum of the angles in triangle *SRT* is 180°, we have:

$$x + 50 + 70 = 180$$
$$x + 120 = 180$$
$$x = 60°.$$

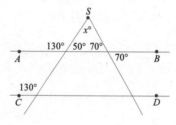

Right

Right triangles contain one right angle. Since the right angle is 90°, the other two angles are complementary. They may or may not be equal to one another. The side of a right triangle opposite the right angle is called the *hypotenuse*. The other two sides are called *legs*. The *Pythagorean theorem* states that the square of the length of the hypotenuse is equal to the sum of the squares of the lengths of the legs.

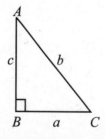

AC is the hypotenuse.
AB and *BC* are legs.
∠*B* = 90°.
∠*A* + ∠*C* = 90°.
$a^2 + c^2 = b^2$.

Example 1

A. If ABC is a right triangle with right angle at B, and if $AB = 6$ and $BC = 8$, what is the length of AC?

$$AB^2 + BC^2 = AC^2$$
$$6^2 + 8^2 = 36 + 64 = 100 = AC^2$$
$$AC = 10$$

B. If the measure of angle A is 30°, what is the measure of angle C?

Since angles A and C are complementary:

$$30° + C = 90°$$
$$C = 60°$$

If the lengths of the three sides of a triangle are a, b, and c and the relation $a^2 + b^2 = c^2$ holds, the triangle is a right triangle and side c is the hypotenuse.

Example 2

Show that a triangle of sides 5, 12, and 13 is a right triangle. The triangle will be a right triangle if $a^2 + b^2 = c^2$.

$$5^2 + 12^2 = 13^2$$
$$25 + 144 = 169$$

Therefore, the triangle is a right triangle and 13 is the length of the hypotenuse.

Example 3

A plane takes off from the airport in Buffalo and flies 600 miles to the north and then flies 800 miles to the east to City C. What is the straight-line distance from Buffalo to City C?

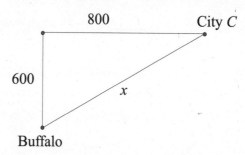

As the diagram above shows, the required distance x is the hypotenuse of the triangle. Thus,

$(600)^2 + (800)^2 = x^2$

$3,600 + 6,400 = x^2$

$10,000 = x^2$

$x = \sqrt{100,000} = 1,000$

Thus, the distance from Buffalo to City C is 1,000 miles.

Area of a Triangle

An *altitude* (or height) of a triangle is a line segment dropped as a perpendicular from any vertex to the opposite side. The area of a triangle is the product of one half the altitude and the base of the triangle. (The base is the side opposite the vertex from which the perpendicular was drawn.)

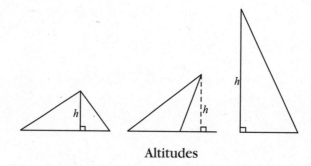

Altitudes

Example 1

What is the area of a right triangle with sides 5, 12, and 13?

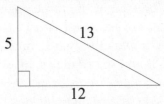

As the picture above shows, the triangle has hypotenuse 13 and legs 5 and 12. Since the legs are perpendicular to one another, we can use one as the height and one as the base of the triangle. Therefore, we have:

$A = \frac{1}{2} \cdot bh$

$A = \frac{1}{2}(12)(5)$

$A = 30$

The area of the triangle is 30.

Example 2

Find the area A of the following isosceles triangle.

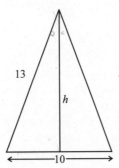

In an isosceles triangle, the altitude from the vertex angle bisects the base (cuts it in half).

The first step is to find the altitude. By the Pythagorean theorem,

$a^2 + b^2 = c^2$; $c = 13$, $a = h$, and $b = \frac{1}{2}(10) = 5$.

$h^2 + 5^2 = 13^2$

$h^2 + 25 = 169$

$h^2 = 144$

$h = 12$

$A = \frac{1}{2} \cdot \text{base} \cdot \text{height}$

$= \frac{1}{2} \cdot 10 \cdot 12$

$= 60$

Similarity

Two triangles are *similar* if all three pairs of corresponding angles are equal. The sum of the three angles of a triangle is 180°; therefore, if two angles of triangle I equal two corresponding angles of triangle II, the third angle of triangle I must be equal to the third angle of triangle II and the triangles are similar. The lengths of the sides of similar triangles are in proportion to each other. A line drawn parallel to one side of a triangle divides the triangle into two portions, one of which is a triangle. The new triangle is similar to the original triangle.

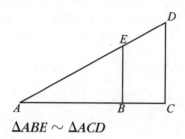

$\triangle ABE \sim \triangle ACD$

Example 1

In the following figure, if $AC = 28$ feet, $AB = 35$ feet, $BC = 21$ feet, and $EC = 12$ feet, find the length of DC if $DE \parallel AB$.

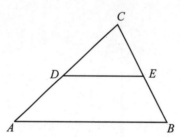

Because $DE \parallel AB$, $\triangle CDE \sim \triangle CAB$. Since the triangles are similar, their sides are in proportion:

$$\frac{DC}{AC} = \frac{EC}{BC}$$

$$\frac{DC}{28} = \frac{12}{21}$$

$$DC = \frac{12 \cdot 28}{21} = 16 \text{ feet}$$

Example 2

A pole that is sticking out of the ground vertically is 10 feet tall and casts a shadow of 6 feet. At the same time, a tree next to the pole casts a shadow of 24 feet. How tall is the tree?

Below is a diagram of the tree and the pole. At the same time of the day, nearby objects and their shadows form similar triangles.

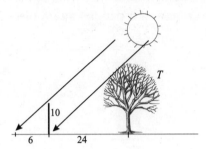

Call the height of the tree T. Then we can write a proportion between the corresponding sides of the triangles.

$$\frac{10}{T} = \frac{6}{24}$$

To solve this proportion, multiply by $24T$.

$$24 \times 10 = 6T$$
$$240 = 6T$$
$$T = 40$$

The tree is 40 feet tall.

Quadrilaterals

A quadrilateral is a polygon of four sides. The sum of the angles of a quadrilateral is 360°. If the opposite sides of a quadrilateral are parallel, the quadrilateral is a *parallelogram*. Opposite sides of a parallelogram are equal and so are opposite angles. Any two consecutive angles of a parallelogram are supplementary. A diagonal of a parallelogram divides the parallelogram into congruent triangles. The diagonals of a parallelogram bisect each other.

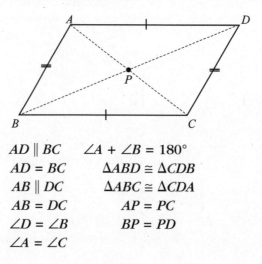

$AD \parallel BC$ $\angle A + \angle B = 180°$
$AD = BC$ $\triangle ABD \cong \triangle CDB$
$AB \parallel DC$ $\triangle ABC \cong \triangle CDA$
$AB = DC$ $AP = PC$
$\angle D = \angle B$ $BP = PD$
$\angle A = \angle C$

Definitions

A *rhombus* is a parallelogram with four equal sides. The diagonals of a rhombus are perpendicular to each other.

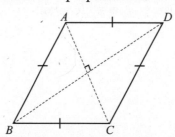

A *rectangle* is a parallelogram with four right angles. The diagonals of a rectangle are equal and can be found using the Pythagorean theorem if the sides of the rectangle are known.

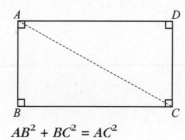

$AB^2 + BC^2 = AC^2$

A *square* is a rectangle with four equal sides.

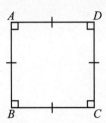

A *trapezoid* is a quadrilateral with only one pair of parallel sides, called *bases*. The nonparallel sides are called *legs*.

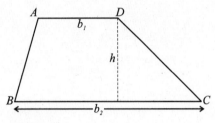

AD ∥ BC
AD and BC are bases
AB and DC are legs
h = altitude

Finding Areas

The area of any *parallelogram* is the product of the base and the height, where the height is the length of an altitude, a line segment drawn from a vertex perpendicular to the base.

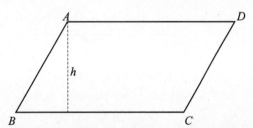

Since rectangles and squares are also parallelograms, their areas follow the same formula. For a *rectangle*, the altitude is one of the sides, and the formula is length times width. Since a *square* is a rectangle for which length and width are the same, the area of a square is the square of its side.

The area of a *trapezoid* is the height times the average of the two bases. The formula is:

$$A = h \frac{b_1 + b_2}{2}$$

The bases are the parallel sides, and the height is the length of an altitude to one of the bases.

Example 1

Find the area of a square whose diagonal is 12 feet. Let s = side of square. By the Pythagorean theorem:

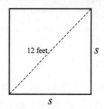

$$s^2 + s^2 = 12^2$$
$$2s^2 = 144$$
$$s^2 = 72$$
$$s = \sqrt{72}$$

Use only the positive value because this is the side of a square.

Since $A = s^2$

$A = 72$ square feet

Example 2

Find the altitude of a rectangle if its area is 320 and its base is 5 times its altitude.

Let altitude = h. Then base = $5h$. Since $A = bh$,

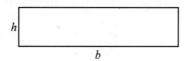

$$A = (5h)(h) = 320$$
$$5h^2 = 320$$
$$h^2 = 64$$
$$h = 8$$

If a quadrilateral is not a parallelogram or trapezoid but is irregularly shaped, its area can be found by dividing it into triangles, attempting to find the area of each, and adding the results.

PLANE GEOMETRY

Example 3

The longer base of a trapezoid is 4 times the shorter base. If the height of the trapezoid is 6 and the area is 75, how long is the longer base?

Recall that the area of a trapezoid is given by the formula

$$A = h \frac{b_1 + b_2}{2}.$$

Let b_1 represent the shorter base. Then the longer base is $b_2 = 4b_1$, and we have

$A = 6 \frac{b_1 + 4b_1}{2} = 6 \frac{5b_1}{2} = 15b$. Since the area is 75, we get

$75 = 15b_1$

$b_1 = 5.$

Thus, the short base is 5 and the long base is 20.

CIRCLES

Definitions

Circles are closed plane curves with all points on the curve equally distant from a fixed point called the *center*. The symbol ⊙ indicates a circle. A circle is usually named by its center. A line segment from the center to any point on the circle is called the *radius* (plural, radii). All radii of the same circle are equal.

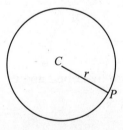

C = center
CP = radius = r

A *chord* is a line segment whose endpoints are on the circle. A *diameter* of a circle is a chord that passes through the center of the circle. A diameter, the longest distance between two points on the circle, is twice the length of the radius. A diameter perpendicular to a chord bisects that chord.

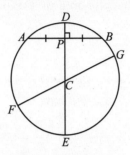

AB is a chord.
C is the center.
DCE is a diameter.
FCG is a diameter.
AB ⊥ DCE so AP = PB.

A *central angle* is an angle whose vertex is the center of a circle and whose sides are radii of the circle. An *inscribed angle* is an angle whose vertex is on the circle and whose sides are chords of the circle.

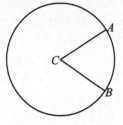 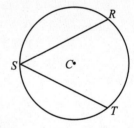

∠ACB is a central angle.
∠RST is an inscribed angle.

An *arc* is a portion of a circle. The symbol ⌒ is used to indicate an arc. Arcs are usually measured in degrees. Since the entire circle is 360°, a semicircle (half a circle) is an arc of 180°, and a quarter of a circle is an arc of 90°.

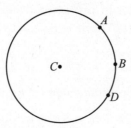

$\overset{\frown}{ABD}$ is an arc.

$\overset{\frown}{AB}$ is an arc.

$\overset{\frown}{BD}$ is an arc.

A central angle is equal in measure to its intercepted arc.

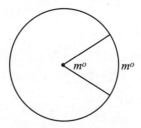

An inscribed angle is equal in measure to one half its intercepted arc. An angle inscribed in a semicircle is a right angle because the semicircle has a measure of 180°, and the measure of the inscribed angle is one half of that.

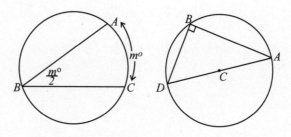

$\overset{\frown}{DA} = 180°$; therefore, $\angle DBA = 90°$.

Perimeter and Area

The perimeter of a circle is called the *circumference*. The length of the circumference is πd, where d is the diameter, or $2\pi r$, where r is the radius. The number π is irrational and can be approximated by 3.14159..., but in problems dealing with circles, it is best to leave π in the answer. There is no fraction exactly equal to π.

Example 1

If the circumference of a circle is 8π feet, what is the radius?

Since $C = 2\pi r = 8\pi$, $r = 4$ feet.

The length of an arc of a circle can be found if the central angle and radius are known. Then the length of the arc is $\frac{n°}{360°}(2\pi r)$, where the central angle of the arc is $n°$. This is true because of the proportion:

$$\frac{\text{Arc}}{\text{Circumference}} = \frac{\text{central angle}}{360°}$$

Example 2

If a circle of radius 3 feet has a central angle of 60°, find the length of the arc intercepted by this central angle.

$$\text{Arc} = \frac{60°}{360°}(2\pi 3) = \pi \text{ feet}$$

The area A of a circle is πr^2, where r is the radius. If the diameter is given instead of the radius,

$$A = \pi \left(\frac{d}{2}\right)^2 = \frac{\pi d^2}{4}.$$

PLANE GEOMETRY

Example 3

Find the area of a circular ring formed by two concentric circles of radii 6 and 8 inches, respectively. (Concentric circles are circles with the same center.)

The area of the ring will equal the area of the large circle minus the area of the small circle.

Area of ring = $\pi 8^2 - \pi 6^2$
$= \pi(64 - 36)$
$= 28\pi$ square inches

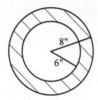

Example 4

A square is inscribed in a circle whose diameter is 10 inches. Find the difference between the area of the circle and that of the square.

If a square is inscribed in a circle, the diagonal of the square is the diameter of the circle. If the diagonal of the square is 10 inches, then, by the Pythagorean theorem,

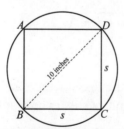

$2s^2 = 100$
$s^2 = 50$

The side of the square s is $\sqrt{50}$, and the area of the square is 50 square inches. If the diameter of the circle is 10, its radius is 5 and the area of the circle is $\pi 5^2 = 25\pi$ square inches. Then the difference between the area of the circle and the area of the square is:

$25\pi - 50$ square inches
$= 25(\pi-2)$ square inches.

Distance Formula

In the arithmetic section, we described the Cartesian coordinate system when explaining how to draw graphs representing linear equations. If two points are plotted in the Cartesian coordinate system, it is useful to know how to find the distance between them. If the two points have coordinates (a, b) and (p, q), the distance between them is:

$$d = \sqrt{(a-p)^2 + (b-q)^2}$$

This formula makes use of the Pythagorean theorem.

Example 1

Find the distance between the two points $(-3, 2)$ and $(1, -1)$.

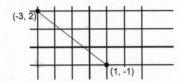

Let $(a, b) = (-3, 2)$ and $(p, q) = (1, -1)$. Then:

$$d = \sqrt{(-3-1)^2 + [2-(-1)]^2}$$
$$= \sqrt{(-4)^2 + (2+1)^2}$$
$$= \sqrt{(-4)^2 + 3^2}$$
$$= \sqrt{16+9} = \sqrt{25} = 5$$

Example 2

What is the area of the circle that passes through the point $(10, 8)$ and has its center at $(2, 2)$?

We can use the distance formula to find the radius of the circle.

$$r = \sqrt{(10-2)^2 + (8-2)^2} = \sqrt{8^2 + 6^2} = \sqrt{100} = 10$$

Thus, the radius of the circle is 10. The area would be $A = \pi r^2 = \pi(10)^2 = 100\pi$.

Three-Dimensional Geometry

Definitions

The volume of any three-dimensional solid figure represents the amount of space contained within it. While area, as we have seen, is measured in square units, the volume of an object is measured in cubic units, such as cubic feet, cubic meters, and cubic centimeters. One cubic foot is defined as the amount of space contained within a cube that is 1 foot on each side.

There are several volume formulas for common solid figures with which you should be familiar.

A rectangular solid is a six-sided figure whose sides are rectangles. The volume of a rectangular solid is its length times its width times its height.

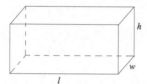

A cube is a rectangular solid whose sides are all the same length. The volume of a cube is the cube of its side.

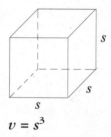

$v = s^3$

The volume of a cylinder is equal to the area of its base times its height. Since the base is a circle, the volume is $V = \pi r^2 h$.

A pyramid has a rectangular base and triangular sides. Its area is given by the formula $V = \frac{1}{3}lwh$.

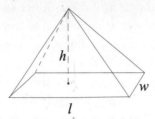

The volume of a cone is given by the formula $V = \frac{1}{3}\pi r^2 h$.

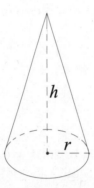

Finally, the formula for the volume of a sphere is given by the formula $V = \frac{4}{3}\pi r^3$.

Example 1

What is the surface area of a cube whose volume is 125 cubic centimeters?

Since the formula for the volume of a cube is $V = s^3$, we have $V = s^3 = 125$. Thus, $s = \sqrt[3]{125} = 5$ centimeters.

If the side of the cube is 5 centimeters, the area of one of its faces is $5^2 = 25$ square centimeters. Since the cube has 6 faces, its surface area is $6 \times 25 = 150$ square centimeters.

Example 2

The volume of a cylinder having a height of 12 is 144π. What is the radius of its base?

The formula for the volume of a cylinder is $V = \pi r^2 h$. Since $V = 144\pi$ and $h = 12$, we have

$144\pi = \pi r^2(12)$.

Divide both sides by π.

$144 = 12r^2$
$12 = r^2$
$r = \sqrt{12} = 2\sqrt{3}$.

Thus, the radius of the base is $2\sqrt{3}$.

QUIZ

GEOMETRY PROBLEMS

1. A chair is 5 feet from one wall of a room and 7 feet from the wall at a right angle to it. How far is the chair from the intersection of the two walls?

2. In triangle XYZ, XZ = YZ. If angle Z has $a°$, how many degrees are there in angle X?

3. In a trapezoid of area 20, the two bases measure 4 and 6. What is the height of the trapezoid?

4. A circle is inscribed in a square whose side is 8. What is the area of the circle, in terms of π?

5. PQ is the diameter of a circle whose center is R. If the coordinates of P are (8, 4) and the coordinates of Q are (4, 8), what are the coordinates of R?

6. The volume of a cube is 64 cubic inches. What is its surface area?

7. The perimeter of scalene triangle EFG is 95. If FG = 20 and EF = 45, what is the measure of EG?

8. In the diagram below, AB is parallel to CD. Find the measures of x and y.

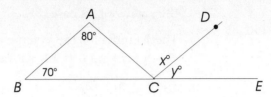

9. In the triangle below, find the measures of the angles.

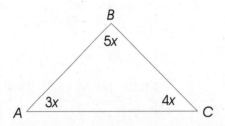

10. If the base of a parallelogram decreases by 20%, and the height increases by 40%, by what percent does the area increase?

11. In the circle below, $AB = 9$ and $BC = 12$. If AC is the diameter of the circle, what is the radius?

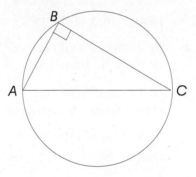

12. In the right triangle below, AB is twice BC. What is the length of BC?

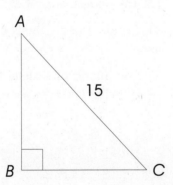

PLANE GEOMETRY

SOLUTIONS

1. As the drawing below shows, we need to find the length of the hypotenuse of a right triangle with legs of 5 and 7. The formula tells us that $5^2 + 7^2 = x^2$, or $74 = x^2$. Thus, $x = \sqrt{74}$.

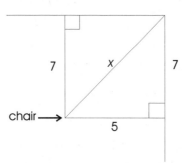

2. The diagram below shows that triangle XYZ is isosceles, and thus angle X and angle Y are the same.

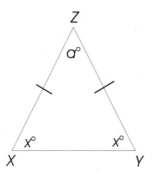

If we call the measure of these angles $x°$, we have $x° + x° + a° = 180°$, or $2x° + a° = 180°$. From this we have $2x° = 180° - a°$ or $x° = \dfrac{(180° - a°)}{2}$.

3. The formula for the area of a trapezoid is $A = \dfrac{1}{2} h (b_1 + b_2)$, where h is the height and b_1 and b_2 are the bases. Substituting, we have $20 = \dfrac{1}{2} h (4 + 6)$ or $40 = 10h$, so $h = 4$.

MATHEMATICS IC AND IIC REVIEW

4. As the picture below shows, the diameter of the circle is 8 and the radius is 4.

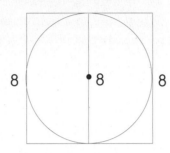

Since $A = \pi r^2$, we have $A = \pi(4^2) = 16\pi$.

5. The center of the circle is the midpoint of the diameter. The formula for the midpoint of a line segment is
$$\left(\frac{(x_1 + x_2)}{2}, \frac{(y_1 + y_2)}{2}\right).$$
Thus, the center is at $\left(\frac{12}{2}, \frac{12}{2}\right) = (6, 6)$.

6. Since the volume of a cube is given by $V = s^3$, where s is the side of the square, it can be seen that s is 4 (since $4^3 = 64$). Then, the surface area of one of the sides is $4 \times 4 = 16$, and, since there are six sides in a cube, the surface area is $16 \times 6 = 96$.

7. $20 + 45 + x = 95$
   $65 + x = 95$
   $x = 30$

8. Since AB and CD are parallel, $\angle BAC$ and $\angle ACD$ are alternate interior angles, and are therefore equal. Thus, $x = 80$. Similarly, $\angle ABC$ is a corresponding angle to $\angle DCE$, and so $y = 70$.

9. Since there are 180° in a triangle, we must have
   $3x + 4x + 5x = 180$
   $12x = 180$
   $x = 15,$

   and $3x = 45$, $4x = 60$, and $5x = 75$. Thus, the angles in the triangle measure 45°, 60°, and 75°.

PLANE GEOMETRY

10. Let b = the length of the base and h = the height in the original parallelogram. Then, the area of the original parallelogram is $A = bh$.

    If the base decreases by 20 percent, it becomes $.8b$. If the height increases by 40 percent, it becomes $1.4h$. The new area, then, is $A = (.8b)(1.4)h = 1.12bh$, which is 12 percent bigger than the original area.

11. Note that triangle ABC is a right triangle. Call the diameter d. Then, we have $9^2 + 12^2 = d^2$, or

    $81 + 144 = d^2$

    $225 = d^2$

    $d = 15.$

    If the diameter is 15, the radius is $7\frac{1}{2}$.

12. Let the length of BC be x. Then, the length of AB is $2x$. By the Pythagorean theorem, we have

    $x^2 + (2x)^2 = 15^2$

    $x^2 + 4x^2 = 225$

    $5x^2 = 225$

    $x^2 = 45$

    $x = \sqrt{45} = \sqrt{9 \times 5}.$

    The length of BC is $3\sqrt{5}$.

COORDINATE GEOMETRY

Both Level IC and Level IIC tests contain coordinate geometry questions.

We have already seen that a coordinate system is an effective way to picture relationships involving two variables. In this section, we will learn more about the study of geometry using coordinate methods.

LINES

Recall that the general equation of a line has the following form:

$$Ax + By + C = 0$$

where A and B are constants and are not both 0. This means that if you were to find all of the points (x, y) that satisfy the above equation, they would all lie on the same line as graphed on a coordinate axis.

If the value of B is not 0, a little algebra can be used to rewrite the equation in the form

$$y = mx + b$$

where m and b are two constants. Since the two numbers m and b determine this line, let's see what their geometric meaning is. First of all, note that the point $(0, b)$ satisfies the above equation. This means that the point $(0, b)$ is one of the points on the line; in other words, the line crosses the y-axis at the point b. For this reason, the number b is called the *y-intercept* of the line.

To interpret the meaning of m, choose any two points on the line. Let us call these points (x_1, y_1) and (x_2, y_2). Both of these points must satisfy the equation of the line above, and so:

$$y_1 = mx_1 + b \text{ and } y_2 = mx_2 + b.$$

If we subtract the first equation from the second, we obtain

$$y_2 - y_1 = m(x_2 - x_1)$$

and solving for m, we find

$$m = (y_2 - y_1)/(x_2 - x_1).$$

The above equation tells us that the number m in the equation $y = mx + b$ is the ratio of the difference of the y-coordinates to the difference of the x-coordinates. This number is called the *slope* of the line. Therefore, the ratio $m = \dfrac{(y_2 - y_1)}{(x_2 - x_1)}$ is a measure of the number of units the line rises (or falls) in the y direction for each unit moved in the x direction. Another way to say this is that the slope of a line is a measure of the rate at which the line rises (or falls). Intuitively, a line with a positive slope rises from left to right; one with a negative slope falls from left to right.

COORDINATE GEOMETRY

Because the equation $y = mx + b$ contains both the slope and the y-intercept, it is called the *slope-intercept* form of the equation of the line.

Example 1

Write the equation $2x - 3y = 6$ in slope-intercept form.

To write the equation in slope-intercept form, we begin by solving for y.

$$-3y = 6 - 2x$$
$$3y = 2x - 6$$
$$y = \frac{2x}{3} - \frac{6}{3} \text{ or}$$
$$y = \frac{2x}{3} - 2$$

Thus, the slope of the line is $\frac{2}{3}$, and the y-intercept is -2.

This, however, is not the only form in which the equation of the line can be written.

If the line contains the point (x_1, y_1), its equation can also be written as:

$$y - y_1 = m(x - x_1).$$

This form of the equation of a line is called the *point-slope* form of the equation of a line, since it contains the slope and the coordinates of one of the points on the line.

Example 2

Write the equation of the line that passes through the point $(2, 3)$ with slope 8 in point-slope form.

In this problem, $m = 8$ and $(x_1, y_1) = (2, 3)$. Substituting into the point-slope form of the equation, we obtain

$$y - 3 = 8(x - 2)$$

Two lines are parallel if and only if they have the same slope. Two lines are perpendicular if and only if their slopes are negative inverses of one another. This means that if a line has a slope m, any line perpendicular to this line must have a slope of $-\frac{1}{m}$. Also note that a horizontal line has a slope of 0. For such a line, the slope-intercept form of the equation reduces to $y = b$.

Finally, note that if $B = 0$ in the equation $Ax + By + C = 0$, the equation simplifies to

$$Ax + C = 0$$

and represents a vertical line (a line parallel to the y-axis) that crosses the x-axis at $-\frac{C}{A}$. Such a line is said to have no slope.

Example 3

Find the slope and the y-intercept of the following lines.

    a. $y = 5x - 7$

    b. $3x + 4y = 5$

a. $y = 5x - 7$ is already in slope-intercept form. The slope is 5, and the y-intercept is -7.

b. Write $3x + 4y = 5$ in slope-intercept form:

$$4y = -3x + 5$$

$$y = \left(-\frac{3}{4}\right)x + \left(\frac{5}{4}\right)$$

The slope is $-\frac{3}{4}$, and the y intercept is $\frac{5}{4}$. This means that the line crosses the y-axis at the point $\frac{5}{4}$, and for every 3 units moved in the x direction, the line falls 4 units in the y direction.

Example 4

Find the equations of the following lines:

    a. the line containing the points (4, 5) and (7,11)

    b. the line containing the point (6, 3) and having slope 2

    c. the line containing the point (5, 2) and parallel to $y = 4x + 7$

    d. the line containing the point $(-2, 8)$ and perpendicular to $y = -2x + 9$

Solutions to Example 4

a. First, we need to determine the slope of the line.

$$m = \frac{(11-5)}{(7-4)} = \frac{6}{3} = 2.$$

Now, using the point-slope form:

$$y - 5 = 2(x - 4).$$

If desired, you can change this to the slope-intercept form: $y = 2x - 3$.

b. Since we know the slope and a point on the line, we can simply plug into the point-slope form:

$$y - 3 = m(x - 6) \text{ to obtain}$$

$$y - 3 = 2(x - 6).$$

c. The line $y = 4x + 7$ has a slope of 4. Thus, the desired line can be written as $y - 2 = 4(x - 5)$.

d. The line $y = -2x + 9$ has a slope of -2. The line perpendicular to this one has a slope of $\frac{1}{2}$. The desired line can be written as $y - 8 = \left(\frac{1}{2}\right)(x + 2)$.

CIRCLES

From a geometric point of view, a circle is the set of points in a plane, each of whose members is the same distance from a particular point called the center of the circle. We can determine the equation of a circle by manipulating the distance formula.

Suppose that we have a circle whose radius is a given positive number r and whose center lies at the point (h, k). If (x, y) is a point on the circle, then its distance from the center of the circle would be

$$\sqrt{(x-h)^2 + (y-k)^2}$$

and since this distance is r, we can say

$$\sqrt{(x-h)^2 + (y-k)^2} = r.$$

Squaring both sides, we get the following result: the equation of a circle whose center is at (h, k) and whose radius is r is given by:

$$(x - h)^2 + (y - k)^2 = r^2$$

Example 1

Find the equation of the circle with radius 7 and center at $(0, -5)$.

Substituting into the formula above, we obtain $x^2 + (y + 5)^2 = 49$.

Example 2

Describe the set of points (x, y) with the property that $x^2 + y^2 > 25$.

The equation $x^2 + y^2 = 25$ describes a circle, centered at the origin, with radius 5. The given set contains all of the points that are *outside* this circle.

QUIZ

COORDINATE GEOMETRY PROBLEMS

1. Find the slope of the line containing the points $(-2, -4)$ and $(2, 4)$.

2. Find the slope of the line given by the equation $4x + 5y = 7$.

3. Find the equation of the line with y-intercept 4 and x-intercept 7.

4. Find the equation of the line through the point $(7, 2)$ and having the same slope as the line through $(2, 4)$ and $(3, -1)$.

5. Find the equation of the line through $(-2, 3)$ and perpendicular to the line $2x - 3y = 4$.

6. What are the center and the radius of the circle given by the equation $(x - 3)^2 + (y + 7)^2 = 81$?

7. Write the equation $4x - 5y = 12$ in slope-intercept form.

8. Find the equation of the line parallel to $x = 7$ and containing the point $(3, 4)$.

9. Write an inequality that represents all of the points inside the circle centered at $(4, 5)$ with radius 4.

10. Find the equation of the line perpendicular to $x = -3$, containing the point $(-3, -6)$.

11. Find the slope of the line containing the points $(-4, 6)$, and $(2, 6)$.

COORDINATE GEOMETRY

SOLUTIONS

1. If (x_1, y_1) and (x_2, y_2) are two points on a line, the slope is given by slope $= \frac{(y_2 - y_1)}{(x_2 - x_1)}$. For the two given points, $(-2, -4)$ and $(2, 4)$, the slope is:

   $$\text{slope} = \frac{(y_2 - y_1)}{(x_2 - x_1)}$$
   $$= \frac{(4 - (-4))}{(2 - (-2))}$$
   $$= \frac{(4 + 4)}{(2 + 2)}$$
   $$= \frac{8}{4} = 2.$$

2. The easiest way to find the slope of the line is to rewrite the equation in the slope-intercept form.

   $4x + 5y = 7$ Subtract $4x$.
   $5y = -4x + 7$ Divide by 5.
   $$y = \left(\frac{-4}{5}\right)x + \left(\frac{7}{5}\right).$$

   The slope of the line is the coefficient of x, that is, $\frac{-4}{5}$.

3. The line has y-intercept 4, which means it passes through $(0,4)$; it also has x-intercept 7, which means it passes through $(7,0)$.

   By the formula, slope $= \frac{(y_2 - y_1)}{(x_2 - x_1)} = \frac{(4-0)}{(0-7)} = \frac{-4}{7}$. Since we know the slope and the y-intercept, we can simply plug into the slope-intercept form, $y = mx + b$, and get $y = \left(\frac{-4}{7}\right)x + 4$.

4. The line through $(2, 4)$ and $(3, -1)$ has slope $\frac{(4 - (-1))}{(2 - 3)} = \frac{5}{(-1)} = -5$. Then, using the point-slope form, the desired line can be written as $y - 2 = -5(x - 7)$.

5. The line $2x - 3y = 4$ can be rewritten as $-3y = -2x + 4$, or $y = \left(\frac{2}{3}\right)x - \left(\frac{4}{3}\right)$. Therefore, its slope is $\frac{2}{3}$, and the line perpendicular to it would have slope $\frac{-3}{2}$. Then, using the point-slope form, the requested line can be written as $y - 3 = \left(\frac{-3}{2}\right)(x + 2)$.

6. The general form for the equation of a circle is $(x - h)^2 + (y - k)^2 = r^2$, where (h, k) is the center and r is the radius. In this case, the given equation can be written as $(x - 3)^2 + (y - (-7))^2 = 9^2$. Therefore, the center is $(3, -7)$ and the radius is 9.

7. To write the equation in slope-intercept form, we begin by solving for y

   $$4x - 5y = 12$$
   $$-5y = -4x + 12$$
   $$y = \frac{-4x}{-5} + \frac{12}{-5} = \frac{4x}{5} - \frac{12}{5}.$$

   Thus, the equation in slope-intercept form is

   $$y = \frac{4x}{5} - \frac{12}{5}.$$

   The slope is $\frac{4}{5}$, and the y-intercept is $-\frac{12}{5}$.

8. Since $x = 7$ is vertical, any line parallel to $x = 7$ will also be vertical. The line parallel to $x = 7$ through $(3, 4)$ is $x = 3$.

9. The equation of the circle with center at $(4, 5)$ with radius 4 is $(x - 4)^2 + (y - 5)^2 = 4^2 = 16$. The points inside this circle are given by the inequality $(x - 4)^2 + (y - 5)^2 < 16$.

10. The line $x = -3$ is vertical, so any line perpendicular to it is horizontal. The horizontal line through the point $(-3, -6)$ is $y = -6$.

11. The slope of the line containing the points $(-4, 6)$ and $(2, 6)$ is $m = \frac{6 - 6}{2 - (-4)} = \frac{0}{6} = 0$. Thus, the line is horizontal.

FUNCTIONS AND THEIR GRAPHS

Both levels of the SAT II Math test contain questions about the elementary functions. The Level II test also contains some questions about more advanced functions (such as trigonometric functions and logarithmic functions), and these are discussed in later sections.

DEFINITIONS

Let D and R be any two sets. Then, we define the *function f from D to R* as a rule that assigns to each member of D one and only one member of R. The set D is called the *domain* of f, and the set R is called the *range* of f. Typically, the letter x is used to represent any element of the domain, and the letter y is used to represent any element of the range.

Note that in the definition above, we have not defined the word "function," but the phrase *function f from D to R*. In order to specify a function, you must not only state the rule, which is symbolized by the letter f, but also the domain D and the range R. However, as you will see, whenever D and R are not specified, there are some generally accepted conventions as to what they are.

In general, the sets D and R can contain any type of members at all. For example, one could define a "telephone number" function in which the domain consists of the names of all of the homeowners in a particular town, the range contains all of the phone numbers in the town, and the rule f associates the homeowners with their phone numbers. However, in mathematics, the sets D and R are usually sets of numbers.

Once again, the letter x is typically used to represent a value of the domain, and the letter y is used to represent a value of the range. Because the value of x determines the value of y (that is, as soon as x is selected, a unique value of y is determined by the rule f), x is called the *independent variable*, and y is called the *dependent variable*.

The symbol $f(x)$, which is read "f of x," is often used instead of y to represent the range value of the function. Often, the rule that specifies a function is expressed in what is called *function notation*. For example, $f(x) = 2x + 3$ specifies a function, which, for each value x in the domain, associates the value $2x + 3$ in the range. For the domain value $x = 7$, we express the corresponding range value as $f(7)$, and compute $f(7) = 2(7) + 3 = 17$. Thus, this function associates the domain value 7 with the range value 17.

Carefully note that the definition of function requires that, for each domain element, the rule associates one and only one range element. This requirement is made to avoid ambiguities when trying to determine the value that f associates to x. For example, consider a domain that once again contains the names of all of the homeowners in a particular town, and let the range be all of the cars in the town. Let the rule f associate each element x in the domain with an element y in the range whenever x is the owner of y. Then, f is *not* a function from D to R, since there could be several homeowners who own more than one car. That is, if x were a homeowner with more than one car, $f(x)$ would not have a well-defined meaning since we wouldn't know which car $f(x)$ actually represents.

Also note that, in general, if a rule f is given and the domain and range are not specified, then the domain of f is assumed to be the set of all real numbers except for those for which $f(x)$ does not exist. The range of f is then the subset of the set of real numbers, which is obtained by plugging all possible values of x into the rule.

Example 1

Consider a rule that assigns the domain values 4, 2, and 12 to the range values 9, 16, and −8 in the following way:

4 → 9

2 → 16

12 → −8

Is this relation between numbers a function?

This relation *is* a function since each element of the domain is assigned to only one element of the range.

Example 2

Consider a rule that assigns the domain values 4, 2, and 12 to the range values 9 and −8 in the following way:

4 → 9

2 ↗

12 → −8

Is this relation between numbers a function?

This relation is also a function. Even though two different domain elements are assigned to the same range element, each element of the domain is assigned to only one element of the range.

FUNCTIONS AND THEIR GRAPHS

Example 3

Consider a rule that assigns the domain values 9 and 3 to the range values 5 and 7 in the following way:

$9 \to 5$

$\nearrow$

$3 \to 7$

Is this relation between numbers a function?

This relation is not a function since the domain value 3 is assigned to two different range values, 5 and 7.

Example 4

If $f(x) = 2x^2 - 5x + 1$, find $f(3), f(0), f(a)$, and $f(b^2)$.

$f(3) = 2(3)^2 - 5(3) + 1 = 18 - 15 + 1 = 4$

$f(0) = 2(0)^2 - 5(0) + 1 = 1$

$f(a) = 2a^2 - 5a + 1$

$f(b^2) = 2(b^2)^2 - 5(b^2) + 1 = 2b^4 - 5b^2 + 1$

Example 5

Find the domain and range of the following functions:

a. $f(x) = 5x^2$

b. $f(x) = \dfrac{1}{x^2}$

For $f(x) = 5x^2$, the domain is the set of all real numbers because the rule produces a real number value for each real value of x. The range is the set of all non-negative real numbers, since, for every value of x, $5x^2 \geq 0$.

For $f(x) = \dfrac{1}{x^2}$, the domain is all real numbers except 0, since the rule is undefined at 0. The range is the set of all positive real numbers since $\dfrac{1}{x^2} > 0$ for all real values of x (except 0).

ARITHMETIC OF FUNCTIONS

Functions can be added, subtracted, multiplied, and divided to form new functions. Let $f(x)$ and $g(x)$ represent two functions. Then, we define the following four functions:

$$(f + g)(x) = f(x) + g(x)$$

This new function is called the *sum* of $f(x)$ and $g(x)$.

$$(f - g)(x) = f(x) - g(x)$$

This new function is called the *difference* of $f(x)$ and $g(x)$.

$$(f \times g)(x) = f(x)g(x)$$

This new function is called the *product* of $f(x)$ and $g(x)$

$$\left(\frac{f}{g}\right)(x) = \frac{f(x)}{g(x)} \text{ where } g(x) \neq 0$$

This new function is called the *quotient* of $f(x)$ and $g(x)$.

Example 1

Let f and g be functions defined by the rules $f(x) = 2x + 3$ and $g(x) = x - 3$. Find $(f + g)(x)$, $(f - g)(x)$, $(f \times g)(x)$, $\left(\frac{f}{g}\right)(x)$, and $\left(\frac{g}{f}\right)(x)$.

$$(f + g)(x) = (2x + 3) + (x - 3) = 3x$$
$$(f - g)(x) = (2x + 3) - (x - 3) = x + 6$$
$$(f \times g)(x) = (2x + 3)(x - 3) = 2x^2 - 3x - 9$$
$$\left(\frac{f}{g}\right)(x) = \frac{2x + 3}{x - 3}$$

Note that the domain of this function is all real numbers except 3.

$$\left(\frac{g}{f}\right)(x) = \frac{x - 3}{2x + 3}$$

Note that the domain is all real numbers except $-\frac{3}{2}$.

It is also possible to combine two functions f and g in a fifth way, known as the *composite function*, written $f \circ g$. The composite function is defined as $f \circ g = f(g(x))$. The composite function can be thought of as representing a chain reaction, where a domain value is first associated with a range value by the rule g, and then this range value is treated as if it were a domain value for f and is associated with a range value for f.

Example 2

If $f(x) = 7x - 5$, find $f(3)$ and $f(0)$.

$f(3) = 7(3) - 5 = 21 - 5 = 16$

$f(0) = -5$

Example 3

Let f and g be functions defined by the rules $f(x) = 3x + 2$ and $g(x) = x - 5$.

Find $(f + g)(x)$, $(f - g)(x)$, $(f \times g)(x)$, $\left(\dfrac{f}{g}\right)(x)$, and $\left(\dfrac{g}{f}\right)(x)$.

$(f + g)(x) = (3x + 2) + (x - 5) = 4x - 3$

$(f - g)(x) = (3x + 2) - (x - 5) = 2x + 7$

$(fg)(x) = (3x + 2)(x - 5) = 3x^2 - 13x - 10$

$\left(\dfrac{f}{g}\right)(x) = \dfrac{(3x + 2)}{(x - 5)}$

Note that the domain of this function is all real numbers except 5.

$\left(\dfrac{g}{f}\right)(x) = \dfrac{(x - 5)}{(3x + 2)}$

Note that the domain is all real numbers except $-\dfrac{2}{3}$.

Example 4

For the functions $f(x) = 2x + 3$ and $g(x) = x - 3$, find $f \circ g$, $g \circ f$, and $f \circ f$.

$f \circ g = f(g(x)) = f(x - 3) = 2(x - 3) + 3 = 2x - 3$

$g \circ f = g(f(x)) = g(2x + 3) = (2x + 3) - 3 = 2x$

$f \circ f = f(f(x)) = f(2x + 3) = 2(2x + 3) + 3 = 4x + 9$

Inverse Functions

The *inverse* of a function, which is written f^{-1}, can be obtained from the rule for the function by interchanging the values of x and y and then solving for y. The inverse of a function will always "undo" the action of a function; for example, if a function f takes the domain value 3 and associates it with 7, then f^{-1} would take the domain value of 7 and associate it with 3.

Example 1

Find the inverse of the function $f(x) = 7x - 4$.

Write $f(x) = 7x - 4$ as $y = 7x - 4$. Then, interchange x and y.

$x = 7y - 4$ Solve for y

$y = \dfrac{(x + 4)}{7}$.   Thus,

$f^{-1}(x) = \dfrac{x + 4}{7}$

Example 2

Find the inverse of the function $f(x) = 3x + 2$.

To begin, write the function as $y = 3x + 2$. Switch x and y to obtain $x = 3y + 2$. Solve for y to obtain $y = \dfrac{(x - 2)}{3}$. Thus,

$f^{-1}(x) = \dfrac{(x - 2)}{3}$.

Intuitively, a function and its inverse "undo" each other. That is, for any function $f(x)$ and its inverse $f^{-1}(x)$, we have $f(f^{-1}(x)) = f^{-1}(f(x)) = x$.

Example 3

Demonstrate that the function $f(x) = 3x + 2$ and its inverse "undo" each other.

Consider $f(x) = 3x + 2$.

As we saw in example 2 above, $f^{-1}(x) = \dfrac{(x - 2)}{3}$.

Then, we have

$f(f^{-1}(x)) = f\left(\dfrac{(x-2)}{3}\right) = 3\left(\dfrac{(x-2)}{3}\right) + 2 = x - 2 + 2 = x.$

Similarly, we can show that $f^{-1}(f(x)) = x$. Thus, since $f(3) = 11$, we have $f^{-1}(11) = 3$.

Graphs of Functions

Functions can be graphed on a coordinate axis by plotting domain values along the *x*-axis and the corresponding *y* values along the *y*-axis. A function of the form $f(x) = c$, where c is any constant, is called a *constant function*, and its graph will always be a horizontal line. A function of the form $f(x) = mx + b$, where m and b are constants, is called a *linear function*. Its graph will always be a straight line that crosses the *y*-axis at b and has a slope of m. (This means that for every unit that the graph runs horizontally, it rises m units vertically.) Finally, a graph of the form

$$f(x) = ax^n + bx^{n-1} + cx^{n-2} + \ldots + kx^2 + mx + p$$

where n is a positive integer, and $a, b, c, \ldots, k, m, p$ are real numbers, is called a *polynomial function*. If $n = 2$, the function looks like $f(x) = ax^2 + bx + c$. Such a function is called a *polynomial function* of degree 2, and its graph is a parabola.

Examples

Graph the following functions: $f(x) = 3x + 4$, $g(x) = x^2 + 2x + 1$.

The function $f(x) = 3x + 4$ is linear, so we only need 2 values to draw the graph. Since $f(0) = 4$, and $f(1) = 7$, we have

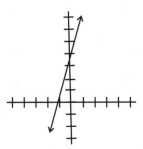

The function $g(x) = x^2 + 2x + 1$ is quadratic, and, thus, the graph will be a parabola. We need to find several points in order to graph the function

$g(0) = 1$
$g(1) = 4$
$g(2) = 9$
$g(-1) = 0$
$g(-2) = 1$
$g(-3) = 4$
$g(-4) = 9$

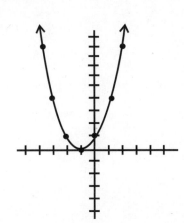

MATHEMATICS IC AND IIC REVIEW

QUIZ

FUNCTIONS PROBLEMS

1. If $f(x) = 3 - 2x$, find $f(-4)$ and $f(11)$.

2. State the domain and range of the following functions:

    a. $f(x) = 3x^2 - 2x + 1$

    b. $g(x) = \dfrac{1}{x}$

3. If $f(x) = x^2$ and $g(x) = 2x - 6$, find the following functions:
    $(f + g)(x)$, $(f \times g)(x)$, $(f \circ g)(x)$, $(f \circ f)(x)$, $g^{-1}(x)$

4. Graph the following functions:

    a. $f(x) = 2x - 3$
    b. $g(x) = x^2 + 2x + 2$

5. If $f(x) = 4 - 3x$, find $f(-5)$ and $f(13)$.

6. What is the domain of the function $y = \sqrt{x + 3}$?

    In the problems below, $f(x) = x^2$ and $g(x) = 4x - 2$.
    Find the following functions:

7. $(f + g)(x)$

8. $(fg)(x)$

9. $(f \circ g)(x)$

10. $(f \circ f)(x)$

11. Find the inverse of $f(x) = 5x + 2$.

SOLUTIONS

1. If $f(x) = 3 - 2x$, $f(-4) = 3 - 2(-4) = 3 + 8 = 11$, and $f(11) = 3 - 2(11) = 3 - 22 = -19$.

2. a. $f(x) = 3x^2 - 2x + 1$

   The domain and range is all real numbers greater than or equal to 2/3.

   b. $g(x) = \dfrac{1}{x}$

   The domain is all real numbers except 0; the range is also all real numbers except 0.

3. If $f(x) = x^2$ and $g(x) = 2x - 6$, $(f + g)(x)$
$$= f(x) + g(x)$$
$$= x^2 + 2x - 6; (f \times g)(x)$$
$$= f(x)g(x)$$
$$= x^2(2x - 6) = 2x^3 - 6x^2; (f \circ g)(x)$$
$$= f(g(x))$$
$$= f(2x - 6) = (2x - 6)^2$$
$$= 4x^2 - 24x + 36; (f \circ f)(x)$$
$$= f(f(x))$$
$$= f(x^2)$$
$$= x^4.$$

   To find $g^{-1}(x)$, write $g(x)$ as $y = 2x - 6$.
   Switch x and y and solve for y.
   $$x = 2y - 6$$
   $$2y = x + 6$$
   $$y = \dfrac{(x + 6)}{2}. \qquad \text{Therefore, } g^{-1}(x) = \dfrac{x + 6}{2}.$$

4. a. Since $f(x)$ is linear, we only need two points to draw the graph. Let's use $f(0) = -3$ and $f(3) = 3$. Then, the graph is:

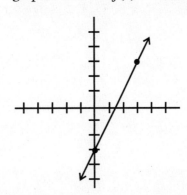

b. $g(x)$ is a quadratic function. We will need to find a series of values until we are able to determine the shape of the graph.

$$g(0) = 2$$
$$g(1) = 5$$
$$g(2) = 10$$
$$g(-1) = 1$$
$$g(-2) = 2$$
$$g(-3) = 5$$
$$g(-4) = 10$$

Drawing the graph, we can see that it is parabolic in shape:

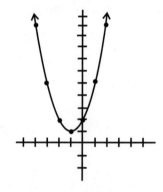

5. $f(-5) = 4 - 3(-5) = 19$

   $f(13) = 4 - 3(13) = 4 - 39 = -35$

6. $y = \sqrt{x + 3}$ The domain is $\{x \mid x \geq -3\}$.

7. $(f + g)(x) = x^2 + 4x - 2$

8. $(fg)(x) = x^2(4x - 2) = 4x^3 - 2x^2$

9. $(f \circ g)(x) = f(4x - 2) = (4x - 2)^2 = 16x^2 - 16x + 4$

10. $(f \circ f)(x) = (x^2)^2 = x^4$

11. Write $f(x) = 5x + 2$ as $y = 5x + 2$. Switch x and y to get $x = 5y + 2$. Solve for y to get $y = \dfrac{(x - 2)}{5}$. Thus, $f^{-1}(x) = \dfrac{(x - 2)}{5}$.

TRIGONOMETRY

There are trigonometry questions on both levels of math tests. However, the trigonometry on Level IIC goes quite a bit beyond what is required for Level IC. Therefore, students who plan to take Level IC need to read only a portion of the following section. The point where those taking IC can stop reviewing and move on to the next section is clearly indicated.

THE TRIGONOMETRIC RATIOS

Every right triangle contains two acute angles. With respect to each of these angles, it is possible to define six ratios, called the trigonometric ratios, each involving the lengths of two of the sides of the triangle. For example, consider the following triangle ABC.

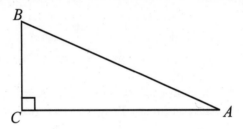

In this triangle, side AC is called the side adjacent to angle A, and side BC is called the side opposite angle A. Similarly, side AC is called the side opposite angle B, and side BC is called the side adjacent to angle B. Of course, side AB is referred to as the hypotenuse with respect to both angles A and B.

The six trigonometric ratios with respect to angle A, along with their standard abbreviations, are given below:

Sine of angle $A = \sin A = \dfrac{\text{opposite}}{\text{hypotenuse}} = \dfrac{BC}{AB}$

Cosine of angle A

$= \cos A = \dfrac{\text{adjacent}}{\text{hypotenuse}} = \dfrac{AC}{AB}$

Tangent of angle $A = \tan A = \dfrac{\text{opposite}}{\text{adjacent}} = \dfrac{BC}{AC}$

Cotangent of angle A

$= \cot A = \dfrac{\text{adjacent}}{\text{opposite}} = \dfrac{AC}{AB}$

Secant of angle A

$= \sec A = \dfrac{\text{hypotenuse}}{\text{adjacent}} = \dfrac{AB}{AC}$

Cosecant of angle A

$$= \csc A = \frac{\text{hypotenuse}}{\text{opposite}} = \frac{AB}{BC}$$

The last three ratios are actually the reciprocals of the first three, in particular:

$$\cot A = \frac{1}{\tan A}$$

$$\sec A = \frac{1}{\cos A}$$

$$\csc A = \frac{1}{\sin A}$$

Also note that:

$$\frac{\sin A}{\cos A} = \tan A, \text{ and } \frac{\cos A}{\sin A} = \cot A.$$

In order to remember which of the trigonometric ratios is which, you can memorize the well-known acronym: **SOH−CAH−TOA**. This stands for: **S**ine is **O**pposite over **H**ypotenuse, **C**osine is **A**djacent over **H**ypotenuse, **T**angent is **O**pposite over **A**djacent.

TRIGONOMETRY

Example 1
Consider right triangle *DEF* below, whose sides have the lengths indicated. Find sin *D*, cos *D*, tan *D*, sin *E*, cos *E*, and tan *E*.

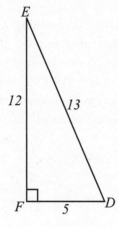

$$\sin D = \frac{EF}{ED} = \frac{12}{13} \quad \sin E = \frac{DF}{ED} = \frac{5}{13}$$

$$\cos D = \frac{DF}{ED} = \frac{5}{13} \quad \cos E = \frac{EF}{ED} = \frac{12}{13}$$

$$\tan D = \frac{EF}{DF} = \frac{12}{5} \quad \tan E = \frac{DF}{EF} = \frac{5}{12}$$

Note that the sine of *D* is equal to the cosine of *E*, and the cosine of *D* is equal to the sine of *E*.

Example 2
In right triangle *ABC*, $\sin A = \frac{4}{5}$. Find the values of the other 5 trigonometric ratios.

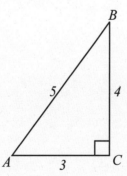

Since the sine of A = opposite over hypotenuse = $\frac{4}{5}$, we know that $BC = 4$, and $AB = 5$. We can use the Pythagorean theorem to determine that $AC = 3$. Then:

$$\cos A = \frac{3}{5}, \tan A = \frac{4}{3}, \cot A = \frac{3}{4}, \sec A = \frac{5}{3}, \csc A = \frac{5}{4}.$$

Trigonometric Ratios for Special Angles

The actual values for the trigonometric ratios for most angles are irrational numbers, whose values can most easily be found by looking in a trigonometry table or using a calculator. There are, however, a few angles whose ratios can be obtained exactly. The ratios for 30°, 45°, and 60° can be determined from the properties of the 30–60–90 right triangle and the 45–45–90 right triangle. First of all, note that the Pythagorean theorem can be used to determine the following side and angle relationships in 30–60–90 and 45–45–90 triangles:

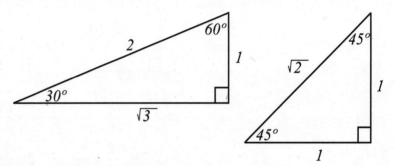

From these diagrams, it is easy to see that:

$$\sin 30° = \frac{1}{2}, \cos 30° = \frac{\sqrt{3}}{2},$$

$$\tan 30° = \frac{1}{\sqrt{3}} = \frac{\sqrt{3}}{3}$$

$$\sin 60° = \frac{\sqrt{3}}{2}, \cos 60° = \frac{1}{2}, \tan 60° = \sqrt{3}$$

$$\sin 45° = \cos 45° = \frac{1}{\sqrt{2}} = \frac{\sqrt{2}}{2}, \tan 45° = 1$$

TRIGONOMETRY

Example

From point A, which is directly across from point B on the opposite sides of the banks of a straight river, the measure of angle CAB to point C, 35 meters upstream from B, is $30°$. How wide is the river?

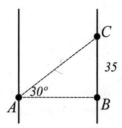

To solve this problem, note that

$$\tan A = \frac{\text{opposite}}{\text{adjacent}} = \frac{BC}{AB} = \frac{35}{AB}.$$

Since the measure of angle A is $30°$, we have

$\tan 30° = \frac{35}{AB}$. Then:

$$AB = \frac{35}{\tan 30°} = \frac{35}{\sqrt{3}/3} = \frac{105}{\sqrt{3}}.$$

Therefore, the width of the river is $\frac{105}{\sqrt{3}}$ meters, or approximately 60 meters wide.

THE PYTHAGOREAN IDENTITIES

There are three fundamental relationships involving the trigonometric ratios that are true for all angles and are helpful when solving problems. They are:

$\sin^2 A + \cos^2 A = 1$
$\tan^2 A + 1 = \sec^2 A$
$\cot^2 A + 1 = \csc^2 A$

These three identities are called the Pythagorean identities since they can be derived from the Pythagorean theorem. For example, in triangle ABC below:

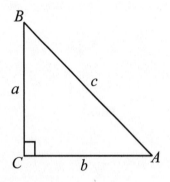

$a^2 + b^2 = c^2$

Dividing by c^2, we obtain,

$\dfrac{a^2}{c^2} + \dfrac{b^2}{c^2} = 1$, or

$\left(\dfrac{a}{c}\right)^2 + \left(\dfrac{b}{c}\right)^2 = 1.$

Now, note that $\dfrac{a}{c} = \sin A$ and $\dfrac{b}{c} = \cos A$. Substituting these values in, we obtain $\sin^2 A + \cos^2 A = 1$. The other two identities are similarly obtained.

Example

If, in triangle ABC, $\sin A = \dfrac{7}{9}$, what are the values of $\cos A$ and $\tan A$?

Using the first of the trigonometric identities, we obtain:

$$\left(\frac{7}{9}\right)^2 + \cos^2 A = 1$$

$$= \frac{49}{81} + \cos^2 A = 1$$

$$\cos^2 A = 1 - \frac{49}{81}$$

$$\cos A = \sqrt{\frac{32}{81}}$$

$$= \frac{4\sqrt{2}}{9}$$

Then, since $\tan A = \dfrac{\sin A}{\cos A}$, we have

$$\tan A = \frac{\left(\dfrac{7}{9}\right)}{\left(\dfrac{4\sqrt{2}}{9}\right)} = \frac{7}{4\sqrt{2}} = \frac{7\sqrt{2}}{8}.$$

QUIZ

TRIGONOMETRY PROBLEMS

1. In right triangle PQR, $\cot P = \dfrac{5}{12}$. Find the value of $\tan P$, $\sin P$, and $\sec P$.

2. Find the value of $\cot 45° + \cos 30° + \sin 150°$.

3. If $\sin a = \dfrac{3}{7}$, and $\cos a < 0$, what is the value of $\tan a$?

4. A wire extends from the top of a 50-foot pole to a stake in the ground. If the wire makes an angle of 55° with the ground, find the length of the wire.

5. A road is inclined at an angle of 10° with the horizontal. If John drives 50 feet up the road, how many feet above the horizontal is he?

6. If $\sin \theta = \dfrac{1}{2}$, and $\cos \theta = -\dfrac{\sqrt{3}}{2}$, find the values of the other 4 trigonometric functions.

7. Demonstrate that $\dfrac{1}{\sin^2 x}$ is equivalent to $\dfrac{(1 + \tan^2 x)}{\tan^2 x}$.

8. In right triangle DEF, with angle F a right angle, $\csc D = \dfrac{13}{12}$. What is the value of $\tan D$?

9. The angle of elevation from an observer at ground level to a vertically ascending rocket measures 55°. If the observer is located 5 miles from the lift-off point of the rocket, what is the altitude of the rocket?

10. What is the value of $3 \cos 45° + 3 \sin 30°$?

SOLUTIONS

1.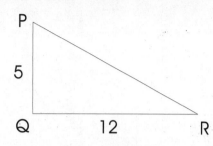

    $\cot P = \dfrac{5}{12} = \dfrac{\text{adjacent}}{\text{opposite}}.$

    Thus, $\tan P = \dfrac{12}{5}.$

    From the Pythagorean theorem, we can compute that the hypotenuse of the triangle is 13.

    Thus, $\sin P = \dfrac{\text{opposite}}{\text{hypotenuse}} = \dfrac{12}{13},$

    and $\sec P = \dfrac{\text{hypotenuse}}{\text{adjacent}} = \dfrac{13}{5}.$

2. Note that $\sin 150° = \sin 30°$. Thus, $\cot 45° + \cos 30° + \sin 150° = 1 + \dfrac{\sqrt{3}}{2} + \dfrac{1}{2} = \dfrac{3 + \sqrt{3}}{2}.$

3. Given $\sin a = \dfrac{3}{7}$, and $\cos a < 0$, we can find the value of $\cos a$ using the Pythagorean identity $\sin^2 a + \cos^2 a = 1.$

    $$\left(\dfrac{3}{7}\right)^2 + \cos^2 a = 1$$

    $$\dfrac{9}{49} + \cos^2 a = 1$$

    $$\cos^2 a = 1 - \dfrac{9}{49} = \dfrac{40}{49}$$

    $$\cos a = -\sqrt{\dfrac{40}{49}} = \dfrac{-2\sqrt{10}}{7}$$

    Since $\tan a = \dfrac{\sin a}{\cos a}$, we have

    $$\tan a = \dfrac{\left(\dfrac{3}{7}\right)}{\dfrac{-2\sqrt{10}}{7}} = \dfrac{-3}{(2\sqrt{10})} = \dfrac{-(3\sqrt{10})}{20}.$$

4. Let L = length of the wire.

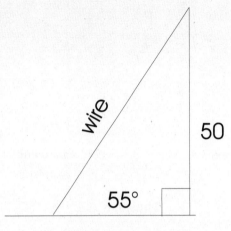

$\sin 55° = \dfrac{50}{L}$, so $L = \dfrac{50}{\sin 55°}$.

5 Let x = the distance above the horizontal.

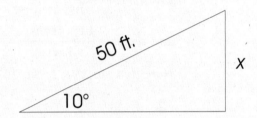

$\sin 10° = \dfrac{x}{50}$. Therefore, $x = 50 \sin 10°$.

6. First of all, we have $\csc \theta = \dfrac{1}{\sin \theta} = 2$.

Similarly, $\sec \theta = \dfrac{1}{\cos \theta} = \dfrac{-2}{\sqrt{3}} = \dfrac{-2\sqrt{3}}{3}$.

Next, $\tan \theta = \dfrac{\sin \theta}{\cos \theta} = \dfrac{\left(\dfrac{1}{2}\right)}{\left(\dfrac{-\sqrt{3}}{2}\right)} = \dfrac{-1}{\sqrt{3}} = -\dfrac{\sqrt{3}}{3}$.

The value of $\cot \theta$ is the reciprocal of $\tan \theta$, which is $-\sqrt{3}$.

7. Solving this problem simply requires the application of several of the fundamental trigonometric identities. First of all, recall that $1 + \tan^2 x = \sec^2 x$. Therefore,

$$\frac{(1 + \tan^2 x)}{\tan^2 x} = \frac{\sec^2 x}{\tan^2 x}.$$

Now, use the fact that $\sec x = \dfrac{1}{\cos x}$ and $\tan x = \dfrac{\sin x}{\cos x}$ to obtain

$$\frac{\sec^2 x}{\tan^2 x} = \frac{\left(\dfrac{1}{\cos^2 x}\right)}{\left(\dfrac{\sin^2 x}{\cos^2 x}\right)} = \frac{1}{\cos^2 x} \cdot \frac{\cos^2 x}{\sin^2 x} = \frac{1}{\sin^2 x}.$$

8. In a right triangle, the csc of an angle is equal to the hypotenuse divided by the side opposite the angle. Since we know $\csc D = \dfrac{13}{12}$, we know that the triangle in question has a hypotenuse of 13, and the side opposite angle D is 12.

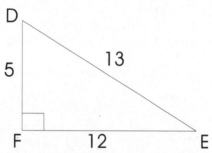

The Pythagorean theorem can be used to determine that the missing (adjacent) side is 5. Now, the tangent of an angle is the ratio of the opposite side to the adjacent side. Therefore,

$$\tan D = \frac{\text{opposite}}{\text{adjacent}} = \frac{12}{5}.$$

9. The best way to begin is with a diagram of the situation:

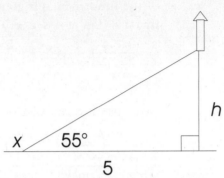

Let h = the altitude of the rocket. Then, since the tangent of an angle is equal to the opposite side divided by the adjacent, we have

$$\tan 55° = \frac{h}{5} \text{ or}$$

$$h = 5 \tan 55°.$$

10. In this problem, we need to know that $\cos 45° = \frac{\sqrt{2}}{2}$ and $\sin 30° = \frac{1}{2}$.

Then,

$$3 \cos 45° + 3 \sin 30° = 3\left(\frac{\sqrt{2}}{2}\right) + 3\left(\frac{1}{2}\right) = \frac{3(\sqrt{2} + 1)}{2}.$$

TRIGONOMETRIC FUNCTIONS OF THE GENERAL ANGLE

The topics in trigonometry that have been covered up to this point are sufficient for those taking the SAT Level I exam. For those taking Level II, however, more knowledge of trigonometry is required. The following sections examine the additional topics in trigonometry that are needed for Level II only.

THE TRIGONOMETRIC DEFINITION OF ANGLES

Previously, in the geometry review section, an angle was defined as two rays meeting at a common endpoint called the vertex. Thus, in geometry, the measure of an angle can range only between 0° and 360°. In trigonometry, angles are defined in a more general way. In the trigonometric definition, an angle is formed as the result of the rotation of a ray.

Consider two rays, one of which, *OA*, is called the *initial side* of the angle, and the other of which, *OB*, is called the *terminal side* of the angle. The angle is said to be in *standard position* if the vertex is at the origin and the initial side is along the positive *x*-axis.

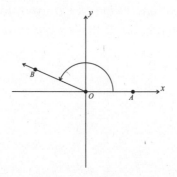

An angle is created by rotating the initial ray *OA* into the terminal ray *OB*. The direction of the rotation is typically indicated by an arrow. An angle is *positive* if the rotation is in a counterclockwise direction and *negative* if the rotation is in a clockwise direction. An angle whose terminal side is on either the *x*- or *y*-axis is called a *quadrantal angle*.

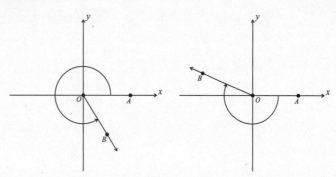

When forming an angle, there is no restriction on the amount of the rotation. The diagram below shows an angle that has been formed by rotating the ray $3\frac{1}{2}$ times.

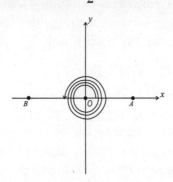

The above angle has its initial side and terminal side in the same position as an angle that has been formed by $2\frac{1}{2}$ rotations, $1\frac{1}{2}$ rotations, or even just $\frac{1}{2}$ of a rotation. It can be seen that many different angles in standard position have the same terminal side. Such angles are called *coterminal*.

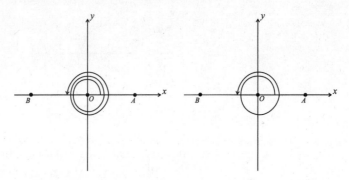

Degrees and Radians

There are two systems for measuring angles. The more common way to measure an angle is in *degrees*. A measure of 360° is assigned to the angle formed by one complete counterclockwise rotation. An angle formed by 3 complete counterclockwise rotations, therefore, would measure 1,080°. An angle formed by one clockwise rotation would have a measure of −360°. Using such a system, it is possible for any real number to be the measure of an angle.

The other system for measuring angles is called the *radians*. As illustrated below, an angle of one radian cuts off from a unit circle (a circle of radius 1 with center at the origin) an arc of length 1

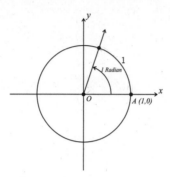

Since the unit circle has a circumference of 2π, one complete rotation is said to measure 2π radians. Thus, an angle of 360° measures 2π radians. Similarly, one half of a rotation (180°) measures π radians, and one quarter of a rotation (a right angle, equal to 90°) measures $\frac{\pi}{2}$ radians. Two complete rotations measures 4π radians, and three complete rotations clockwise measures −6π radians. In general, the radian measure of an angle x is the arc length along the unit circle covered in rotating from the initial side to the terminal side.

The diagrams below show several rotations and the resulting angles in radians.

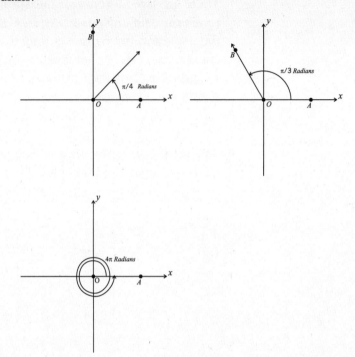

Now, as we have seen that 360° = 2π radians, it follows that 1° = $\frac{\pi}{180}$ radians, and 1 radian = $\frac{180°}{\pi}$. The ratio $\frac{\pi}{180}$, thus, can be used to convert degrees to radians, and the ratio $\frac{180}{\pi}$ can be used to convert radians to degrees.

Example

a. Express 30° in radians.

$$30° = 30\left(\frac{\pi}{180}\right) = \frac{\pi}{6} \text{ radians}$$

b. Express the angle of $\frac{\pi}{10}$ radians in degrees.

$$\frac{\pi}{10} = \frac{\pi}{10}\left(\frac{180}{\pi}\right) = 18°$$

THE TRIGONOMETRIC FUNCTIONS

The six trigonometric ratios previously defined can be defined as functions in the following way: Let θ represent a trigonometric angle in standard position. Further, let $P(x, y)$ represent the point of intersection between the terminal side of the angle and the unit circle.

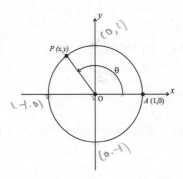

Then, the sine and cosine functions are defined as follows:

$\sin \theta = y$

$\cos \theta = x$

$\sin \theta = y$ means that the sine of θ is the directed vertical distance from the y-coordinate of $P(x, y)$ to the x-axis. Thus, as θ progresses from 0 radians to $\frac{\pi}{2}$ radians to π radians to $\frac{3\pi}{2}$ radians to 2π radians, the value of sin θ goes from 0 to 1 back to 0, to −1 and back to 0. Further, as θ progresses beyond 2π radians, the values of sin θ repeat in the same pattern: 0 to 1 to 0 to −1 to 0. Thus, we say that the sine function is periodic with period 2π. Another way to view this is to realize that sin 0 = 0, $\sin\left(\frac{\pi}{2}\right) = 1$, sin π = 0, $\sin\left(\frac{3\pi}{2}\right) = -1$, sin 2π = 0, and so on.

$\cos \theta = x$ means that the cosine of θ is the directed horizontal distance from the x-coordinate of $P(x, y)$ to the y-axis. Thus, as θ progresses from 0 radians to $\frac{\pi}{2}$ radians to π radians to $\frac{3\pi}{2}$ radians to 2π radians, the value of cos θ goes from 1 to 0 to −1, back to 0, and then back to 1. Then, as θ progresses beyond 2π radians, the values of cos θ repeat in the same pattern: 1 to 0 to −1 to 0 to 1. Clearly, then, cosine is also periodic with period 2π. This means that cos 0 = 1, $\cos\left(\frac{\pi}{2}\right) = 0$, cos π = −1, $\cos\left(\frac{3\pi}{2}\right) = 0$, cos 2π = 1, and so on.

After defining sine and cosine, we can define four more trigonometric functions, starting with the tangent function, which is defined as $\tan \theta = \frac{y}{x}$. Note that $\tan 0 = 0$ since $y = 0$ for an angle of 0 radians. Next, $\tan \frac{\pi}{2} = \frac{1}{0}$ and is, thus, undefined. The value for $\tan \pi$ is $\frac{0}{-1} = 0$, $\tan\left(\frac{3\pi}{2}\right) = \frac{-1}{0}$ and is, once again, undefined. Finally, $\tan 2\pi = 0$.

The other three trigonometric functions are defined as the reciprocals of the three functions already defined. The cosecant function, thus, is $\csc \theta = \frac{1}{\sin \theta}$, defined for all θ such that $\sin \theta \neq 0$. The secant function is defined as $\sec \theta = \frac{1}{\cos \theta}$, defined for all θ such that $\cos \theta \neq 0$. Finally, the cotangent function is defined as $\cot \theta = \frac{1}{\tan \theta}$, defined for all θ such that $\tan \theta \neq 0$.

TRIGONOMETRIC FUNCTIONS OF THE GENERAL ANGLE

REFERENCE ANGLES AND FINDING TRIGONOMETRIC FUNCTIONS OF ANY ANGLE

For any angle θ not in the first quadrant, the values of the trigonometric functions of θ can be expressed in terms of a first quadrant reference angle. For any angle θ (except $\frac{\pi}{2}$ and $\frac{3\pi}{2}$), the measure of the reference angle θ' in the first quadrant is equal to the measure of the acute angle formed by the terminal side of θ and the x-axis. The figure below depicts the reference angles for θ with terminal side in each of the quadrants II, III, and IV, respectively.

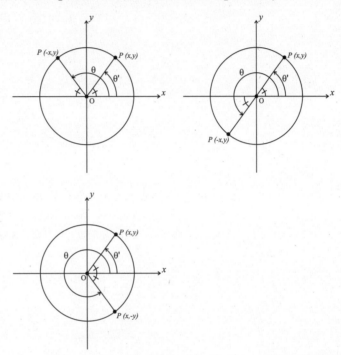

The value of a trigonometric function for a particular angle in the second, third, or fourth quadrant is equal to either plus or minus the value of the function for the first quadrant reference angle. The sign of the value is dependent upon the quadrant that the angle is in.

Example 1

Express each of the following in terms of first quadrant reference angles.

a. sin 340°

b. cos 170°

a. The angle 340° is in the fourth quadrant, and, as the picture below indicates, has a reference angle of 20°. Since sine is negative in the fourth quadrant, sin 340° = −sin 20°.

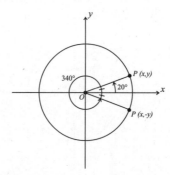

b. The angle 170° is in the second quadrant, and, as the picture below indicates, has a reference angle of 10°. Since cosine is negative in the second quadrant, cos 170° = −cos 10°.

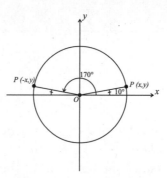

We have already seen that the sine and cosine functions repeat their values every 2π units. It is said that sine and cosine are periodic with period 2π or period 360°.

Similarly, the tangent and cotangent functions can be shown to be periodic with period π or 180°, and the secant and cosecant functions can be shown to be periodic with period 2π (that is, 360°).

The concept of periodicity can be used along with reference angles to express the value of a trigonometric function of any angle in terms of a first quadrant angle.

Example 2

Express each of the following in terms of first quadrant reference angles.

a. $\sin 425°$

b. $\cos\left(\dfrac{13\pi}{3}\right)$

a. Since $425° = 360° + 65°$, $\cos 425° = \cos 65°$.

b. Since $\dfrac{13\pi}{3} = 4\pi + \dfrac{\pi}{3}$, $\cos\left(\dfrac{13\pi}{3}\right) = \cos\dfrac{\pi}{3}$.

QUIZ

TRIGONOMETRIC FUNCTIONS PROBLEMS

1. Change the following degree measures to radian measure:

   a. $45°$

   b. $-150°$

2. Change the following radian measures to degree measure:

   a. $\dfrac{3\pi}{2}$

   b. $-\dfrac{2\pi}{3}$

3. Find a positive angle of less than one revolution that is coterminal with the following angles:

   a. $400°$

   b. $\dfrac{19\pi}{4}$

4. Express the following angles in terms of first-quadrant reference angles:

   a. $\tan 336°$

   b. $\sec \dfrac{2\pi}{5}$

5. Use the periodicity of the trigonometric functions to find the exact value of the following:

   a. $\tan 21\pi$

   b. $\sin 420°$

SOLUTIONS

1. a. $45° = 45\left(\dfrac{\pi}{180}\right) = \dfrac{\pi}{4}$ radians

    b. $-150° = -150\left(\dfrac{\pi}{180}\right) = -\dfrac{5\pi}{6}$ radians

2. a. $\dfrac{3\pi}{2} = \dfrac{3\pi}{2}\left(\dfrac{180}{\pi}\right) = 270°$

    b. $-\dfrac{2\pi}{3} = -\dfrac{2\pi}{3}\left(\dfrac{180}{\pi}\right) = -120°$

3. a. $400° = 360° + 40°$. Thus, a 40° angle is coterminal with a 400° angle.

    b. $\dfrac{19\pi}{4} = 4\pi + \dfrac{3\pi}{4}$, which is coterminal with an angle of $\dfrac{3\pi}{4}$ radians.

4. a. The reference angle for an angle of 336° is $360° - 336° = 24°$. Thus, $\tan 336° = \pm \tan 24°$. Finally, since tangent is negative in the first quadrant, we have $\tan 336° = -\tan 24$.

    b. Note that $\dfrac{2\pi}{5}$ is already a first-quadrant angle. Thus, $\sec \dfrac{2\pi}{5}$ is already expressed in terms of a first quadrant angle.

5. a. Since the tangent function has a period of π, $\tan 21\pi = \tan 0 = 0$.

    b. $\sin 420° = \sin(360° + 60°) = \sin 60° = \dfrac{\sqrt{3}}{2}$.

Trigonometric Identities

An *identity* is an equation that is true for all possible values of its variables. As an example, the equation $5(x + 4) = 5x + 20$ is an identity, true for all values of x.

While it is fairly obvious that the equation above is an identity, frequently in trigonometry, it is not as easy to tell by inspection whether or not an equation is an identity. ***Proving a trigonometric identity*** refers to the act of demonstrating that a given trigonometric equation is, in fact, an identity. Typically, proving an identity takes a number of steps, and it is accomplished by manipulating one side of the given equation to look exactly like the other side.

When attempting to prove an identity, it is often helpful to use the eight fundamental trigonometric identities, which have already been discussed in this section. For reference purposes, they are repeated below:

The Reciprocal Identities

$$\csc x = \frac{1}{\sin x}$$

$$\sec x = \frac{1}{\cos x}$$

$$\cot x = \frac{1}{\tan x}$$

The Quotient Identities

$$\tan x = \frac{\sin x}{\cos x}$$

$$\cot x = \frac{\cos x}{\sin x}$$

The Pythagorean Identities

$$\sin^2 x + \cos^2 x = 1$$
$$1 + \tan^2 x = \sec^2 x$$
$$1 + \cot^2 x = \csc^2 x$$

Example 1

Verify that $\csc x \cos x = \cot x$ is an identity.

By using the reciprocal identity for csc, we see that

$$\csc x \cos x = \left(\frac{1}{\sin x}\right) \cos x = \frac{\cos x}{\sin x} = \cot x.$$

Example 2

Prove the following identity: $\cos^2 x - \sin^2 x = 2\cos^2 x - 1$.

By using the Pythagorean identity for sin and cos, rewritten in the form $\sin^2 x = 1 - \cos^2 x$, we see that

$$\cos^2 x - \sin^2 x = \cos^2 x - (1 - \cos^2 x)$$
$$= \cos^2 x - 1 + \cos^2 x$$
$$= 2\cos^2 x - 1.$$

Often, the process of proving an identity involves the application of several of the fundamental identities and some arithmetic manipulation.

Example 3

Prove the following identity: $\dfrac{(1 - \cos x)}{(1 + \cos x)} = (\csc x - \cot x)^2$.

$$\frac{(1 - \cos x)}{(1 + \cos x)} = \frac{(1 - \cos x)(1 - \cos x)}{(1 + \cos x)(1 - \cos x)}$$
$$= \frac{(1 - \cos x)^2}{1 - \cos^2 x}$$
$$= \frac{(1 - \cos x)^2}{\sin^2 x}$$
$$= \left(\frac{1 - \cos x}{\sin x}\right)^2$$
$$= \left(\frac{1}{\sin x} - \frac{\cos x}{\sin x}\right)^2$$
$$= (\csc x - \cot x)^2$$

TRIGONOMETRIC FUNCTIONS OF THE GENERAL ANGLE

Additional Trigonometric Identities

Sums and Differences of Angles

Sometimes, when working with trigonometric functions, it becomes necessary to work with functions of the sums and differences of two angles, such as $\sin(x + y)$. These functions can be best handled by expanding them according to the sum and difference formulas below:

$$\sin(x + y) = \sin x \cos y + \cos x \sin y$$
$$\sin(x - y) = \sin x \cos y - \cos x \sin y$$
$$\cos(x + y) = \cos x \cos y - \sin x \sin y$$
$$\cos(x - y) = \cos x \cos y + \sin x \sin y$$
$$\tan(x + y) = \frac{(\tan x + \tan y)}{(1 - \tan x \tan y)}$$
$$\tan(x - y) = \frac{(\tan x - \tan y)}{(1 + \tan x \tan y)}$$

Example

Use one of the formulas above to prove that $\cos(-x) = \cos x$.

$\cos(-x) = \cos(0 - x) = \cos 0° \cos x + \sin 0° \sin x$.

Recall that $\sin 0° = 0$ and $\cos 0° = 1$. Thus, $\cos 0° \cos x + \sin 0° \sin x = 1 \cos x + 0 \sin x = \cos x$.

Double and Half Angle Formulas

The sum and difference formulas can be used to come up with the double angle formulas:

$$\sin 2x = 2 \sin x \cos x$$
$$\cos 2x = \cos^2 x - \sin^2 x$$
$$\tan 2x = \frac{(2 \tan x)}{(1 - \tan^2 x)}$$

By using the Pythagorean identity $\sin^2 x + \cos^2 x = 1$, the following two equivalent expressions for $\cos 2x$ can be obtained:

$$\cos 2x = 1 - 2 \sin^2 x$$
$$\cos 2x = 2 \cos^2 x - 1$$

In a similar way, the double-angle formulas can be used to derive the formulas for the sine, cosine, and tangent of half angles:

$$\sin\left(\frac{x}{2}\right) = \pm\sqrt{\frac{1-\cos x}{2}}$$

$$\cos\left(\frac{x}{2}\right) = \pm\sqrt{\frac{1+\cos x}{2}}$$

$$\tan\left(\frac{x}{2}\right) = \pm\sqrt{\frac{1-\cos x}{1+\cos x}}$$

Note the $\pm$ sign in front of the radicals. It is there because the functions of half angles may be either positive or negative.

Example 1

Use an appropriate sum or difference formula to prove
$$\tan 2x = \frac{(2\tan x)}{(1-\tan^2 x)}.$$

Since, $\tan(x+y) = \frac{(\tan x + \tan y)}{(1 - \tan x \tan y)}$, we have

$$\tan 2x = \tan(x+x) = \frac{(\tan x + \tan x)}{(1 - \tan x \tan x)} = \frac{(2\tan x)}{(1-\tan^2 x)}.$$

Example 2

Given that $\cos 2x = \cos^2 x - \sin^2 x$, prove that $\cos 2x = 1 - 2\sin^2 x$.

Since $\sin^2 x + \cos^2 x = 1$, we have $\cos^2 x = 1 - \sin^2 x$. Thus,

$$\cos 2x = \cos^2 x - \sin^2 x = (1 - \sin^2 x) - \sin^2 x = 1 - 2\sin^2 x.$$

TRIGONOMETRIC EQUATIONS

Unlike identities, a *trigonometric equation* is one that is true for only certain values of the angle in the equation. All of the techniques that have been previously used in the algebra section for solving equations can be, and are, used for solving trigonometric equations.

Since all of the trigonometric functions are periodic, they tend to have multiple solutions. Sometimes, we are looking for all of the solutions only within a certain range, such as between 0° and 360°, and sometimes we are looking for *all* of the solutions. For example, the equation $\sin x = 1$ has one solution, 90°, between 0° and 360°. However, since the sine function is periodic with period 360°, there are other solutions outside of the 0° to 360° range. For example, 90° + 360° = 450° is a solution, as is 90° + 2(360°) = 810°, etc.

In the three examples below, find all of the solutions of the given equations in the 0° to 360° range.

Example 1

$8 \sin x - 6 = 2 \sin x - 3$

Begin by moving the terms involving sin to the left-hand side and the constant terms to the right-hand side of the equation.

$6 \sin x = 3$. Divide by 6.

$\sin x = \frac{1}{2}$. Now, since $\sin x = \frac{1}{2}$ twice in the 0° to 360° range, this equation has two solutions, 30° and 150°.

Example 2

$2 \sin^2 x + \sin x = 1$

This equation is a quadratic equation in sin x and can be solved by factoring:

$2 \sin^2 x + \sin x - 1 = 0$

$(2 \sin x - 1)(\sin x + 1) = 0$

Thus, the equation is solved whenever $2 \sin x = 1$ or $\sin x = -1$. Based on example 1 above, $2 \sin x = 1$ at 30° and 150°. Also, $\sin x = -1$ at 270°. Thus, this equation has 3 solutions in the 0° to 360° range: 30°, 150°, and 270°.

Example 3

$4 \sin^3 x - \sin x = 0$

This equation is a cubic equation in sin x. However, by factoring, we can reduce it to the product of two equations in sin x, one linear and one quadratic:

$4 \sin^3 x - \sin x = \sin x(4 \sin^2 x - 1) = 0$.

Thus, the equation is true whenever $\sin x = 0$ or whenever $4 \sin^2 x = 1$. Now, $\sin x = 0$ at 0° and 180°. In addition, $4 \sin^2 x - 1 = 0$ when $4 \sin^2 x = 1$ or $\sin^2 x = \frac{1}{4}$ or $\sin x = \pm\frac{1}{2}$. This happens at 30°, 150°, 210°, and 330°. Thus, this particular equation has 6 solutions.

MATHEMATICS IC AND IIC REVIEW

QUIZ

TRIGONOMETRIC IDENTITY AND EQUATION PROBLEMS

Verify the following identities:

1. $\csc x - \cos x \cot x = \sin x$
2. $\sin^3 x = \sin x - \sin x \cos^2 x$
3. $\dfrac{\cos x}{1 + \sin x} = \sec x - \tan x$

Solve the following equations in the interval 0° to 360°:

4. $2 \tan^2 x \cos x = \tan^2 x$
5. $\sin^2 2x = \sin 2x$
6. $2\cos^2 x + 3 \cos x + 1 = 0$

SOLUTIONS

1. $\csc x - \cos x \cot x = 1/\sin x - \cos x \left(\dfrac{\cos x}{\sin x}\right)$

$$= \dfrac{1}{\sin x} - \dfrac{\cos^2 x}{\sin x}$$

$$= \dfrac{1 - \cos^2 x}{\sin x}$$

$$= \dfrac{\sin^2 x}{\sin x} = \sin x$$

2. $\sin^3 x = \sin x (\sin^2 x) = \sin x (1 - \cos^2 x) = \sin x - \sin x \cos^2 x$

3. $\dfrac{\cos x}{1 + \sin x} = \dfrac{\cos x}{1 + \sin x} \times \dfrac{1 - \sin x}{1 - \sin x}$

$$= \dfrac{\cos x (1 - \sin x)}{1 - \sin^2 x}$$

$$= \dfrac{\cos x (1 - \sin x)}{\cos^2 x}$$

$$= \dfrac{1 - \sin x}{\cos x}$$

$$= \dfrac{1}{\cos x} - \dfrac{\sin x}{\cos x}$$

$$= \sec x - \tan$$

4. $2\tan^2 x \cos x = \tan^2 x$
$2\tan^2 x \cos x - \tan^2 x = 0$
$\tan^2 x(2\cos x - 1) = 0$

The solutions to $\tan^2 x = 0$ are $0°$ and $180°$.
The solutions to $2\cos x - 1 = 0$ are $60°$ and $300°$.

5. $\sin^2 2x = \sin 2x$
$\sin^2 2x - \sin 2x = 0$
$\sin 2x(\sin 2x - 1) = 0$

The solutions to $\sin 2x = 0$ are $0, 90°, 180°,$ and $270°$.

The solutions to $\sin 2x - 1 = 0$, or $\sin 2x = 1$, are $45°$ and $135°$.

6. $2\cos^2 x + 3\cos x + 1 = 0$

Solve by treating as a quadratic in $\cos x$ and factoring.
$2\cos^2 x + 3\cos x + 1 = 0$
$(2\cos x + 1)(\cos x + 1) = 0$

This equation, thus, is true when $\cos x = -1$ and when $\cos x = -\frac{1}{2}$. The solution to $\cos x = -1$ is $180°$, and the solutions to $\cos x = -\frac{1}{2}$ are $120°$ and $240°$.

SOLVING TRIANGLES

In many mathematical word problems and real-world applications, we are given some information about the sides and angles of a particular triangle and asked to determine the lengths of the sides and measures of the angles that were not given. The process of solving a triangle involves taking the given information about the sides and angles of the triangle and computing the lengths and measures of the missing sides and angles. Previously, we learned how to use the Pythagorean theorem and the basic trigonometric functions to solve *right* triangles. It is also possible to solve triangles that are not right.

The Law of Sines and Law of Cosines are two trigonometric formulas that, together, enable you to solve *any* triangle, as long as enough information has been given to determine the missing sides and angles uniquely.

The Law of Sines tells us that, in any triangle, the lengths of the sides are proportional to the sines of the opposite angles. In other words, in any triangle XYZ with corresponding sides x, y, and z, it is true that

$$\frac{x}{\sin X} = \frac{y}{\sin Y} = \frac{z}{\sin Z}$$

The Law of Cosines is a generalized form of the Pythagorean theorem, differing only in that it has an extra term at the end. The Law of Cosines states that in any triangle XYZ, with corresponding sides x, y, and z,

$$x^2 = y^2 + z^2 - 2yz \cos X$$
$$y^2 = x^2 + z^2 - 2xz \cos Y$$
$$z^2 = x^2 + y^2 - 2xy \cos Z$$

As the following examples show, the Law of Cosines can be used to solve triangles in which we are given the lengths of all three sides or two sides and an included angle. The Law of Sines can be used to solve triangles in which we are given two angles and a side or two sides and the angle opposite one of the sides.

The case in which we are given two sides and the angle opposite one of the sides is called the "ambiguous case" since the triangle may have no solutions, one solution, or two solutions, depending on the values given. In the *AAA* case (given three angles and no sides), the triangle cannot be solved, since there are an infinite number of similar triangles for any given three angles.

Note that since the values of most trigonometric functions are irrational numbers, it will be necessary to use either a set of trigonometry tables or a scientific calculator to solve the problems below. If, on the test, you are instructed not to use a calculator, your answer will need to be written in terms of trigonometry functions such as sin 33.5° or cos 23.8°, etc.

Example 1

Solve the triangle with $A = 50°$, $C = 33.5°$, and $b = 76$.

First, we use the fact that the angles in a triangle add up to 180° to find the value of B: $B = 180° - 50° - 33.5° = 96.5°$.

Then, by the Law of Sines, $\frac{76}{\sin 96.5°} = \frac{a}{\sin 50°}$.

Thus, $a = \left(\frac{76}{\sin 96.5°}\right) \times \sin 50° \approx \left(\frac{76}{.9936}\right) \times .7660 = 58.59$.

Now, to find c, we use the Law of Sines again:

$\frac{76}{\sin 96.5°} = \frac{c}{\sin 33.5°}$.

Thus, $c = \left(\frac{76}{\sin 96.5°}\right) \times \sin 33.5° \approx \left(\frac{76}{.9936}\right) \times .5519 = 42.21$.

Example 2

Find the measures of the three angles in the triangle with sides $a = 5$, $b = 7$, and $c = 10$.

To find angle A, use the formula $a^2 = b^2 + c^2 - 2bc \cos A$. We have

$$5^2 = 7^2 + 10^2 - 2(7)(10)\cos A$$
$$25 = 49 + 100 - 140 \cos A$$
$$-124 = -140 \cos A$$
$$\cos A = \frac{124}{140} \approx .8857.$$

Note that this is the first time that we have been asked to find the measure of a particular angle with a given cosine; that is, we need to find the angle whose cosine is .8857. This is particularly easy to do with a scientific calculator. Simply enter .8857, and then press the key that looks like this: $\cos^{-1}$. This tells the calculator to find the angle whose cosine is .8857. When we do this, we find that $A = 27.66°$.

To find angle B, use the formula $b^2 = a^2 + c^2 - 2ac \cos B$. We have

$$7^2 = 5^2 + 10^2 - 2(5)(10)\cos B$$
$$49 = 25 + 100 - 100 \cos B$$
$$-76 = -100 \cos B$$
$$\cos B = \frac{76}{100} = .7600.$$

From a table or a calculator, we find that $B = 40.53°$.

Finally, by subtracting, we obtain $C = 111.81°$.

Trigonometry enables you to solve problems that involve finding measures of unknown lengths and angles.

SCALARS AND VECTORS

It is possible that, on your test, you might be asked to solve a problem involving vectors. These problems can be solved by using either the right triangle trigonometry already discussed or, if necessary, the Laws of Sines and Cosines already discussed in this section. The section below contains the information that you need to know about vectors.

DEFINITIONS

There are two kinds of physical quantities that are dealt with extensively in science and mathematics. One quantity has magnitude only, and the other has magnitude and direction. A quantity that has magnitude only is called a *scalar* quantity. The length of an object expressed in a particular unit of length, mass, time, and density are all examples of scalars. A quantity that has both magnitude and direction is called a *vector*. Forces, velocities, and accelerations are examples of vectors.

It is customary to represent a vector by an arrow. The length of the arrow represents the magnitude of the vector, and the direction in which the arrow is pointing represents the direction. Thus, a force, for example, could be represented graphically by an arrow pointing in the direction in which the force acts and having a length (in some convenient unit of measure) equal to the magnitude of the force. The vector below, for example, represents a force of magnitude three units, acting in a direction 45° above the horizontal.

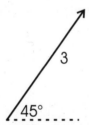

Two vectors are said to be equal if they are parallel, have the same magnitude (length), and point in the same direction. Thus, the two vectors V and U in the diagram below are equal. If a vector has the same magnitude as U but points in the opposite direction, it is denoted as $-U$.

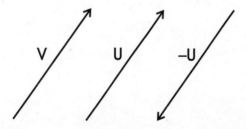

OPERATIONS ON VECTORS

To find the sum of two vectors A and B, we draw from the head of vector A a vector equal to B. The sum of A and B is then defined as the vector drawn from the foot of A to the head of B.

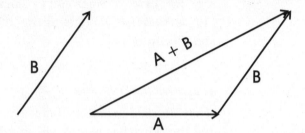

This technique of vector addition can enable us to compute the net effect of two different forces applied simultaneously to the same body.

Example

Two forces, one of magnitude $3\sqrt{3}$ pointing to the east and one of magnitude 3 pointing to the north, act on a body at the same time. Determine the direction in which the body will move and the magnitude of the force with which it will move.

Begin by drawing vectors OA and OB, representing the two forces. Redraw vector OB on the tip of OA. Then, draw in the vector OC, which represents the sum of the two vectors. The body will move in the direction in which this vector is pointing, with a force equal to the magnitude of the vector.

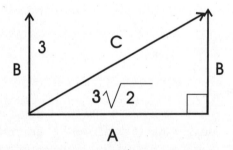

Since OA and OB operate at right angles to each other, we can use the Pythagorean theorem to determine the magnitude of the *resultant* vector.

$$(3\sqrt{2})^2 + 3^2 = C^2$$
$$27 + 9 = C^2$$
$$36 = C^2$$
$$6 = C$$

Further, by recalling the properties of the 30-60-90 triangle, we can see that the resultant force is 30° to the horizontal. Thus, the body will move with a force of 6 units at an angle of 30° to the horizontal.

To subtract the vector U from the vector V, we first draw the vectors from a common origin. Then, the vector extending from the tip of U to the tip of V and pointing to the tip of V is defined as the difference $V - U$.

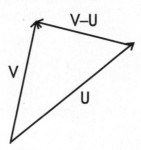

Of course, if the triangle in the problem above had not been a right triangle, it would have been necessary to use the Laws of Sines and Cosines to solve the problem.

QUIZ

TRIANGLE PROBLEMS

In the six problems below, determine the missing parts of triangle ABC from the given information.

1. $A = 35°, B = 25°, c = 67.6$
2. $A = 40°, a = 20, b = 15$
3. $A = 63°, a = 10, c = 8.9$
4. $b = 29, c = 17, A = 103°$
5. $a = 7, b = 24, c = 26$
6. $A = 18°, B = 47°, C = 115°$

MATHEMATICS IC AND IIC REVIEW

SOLUTIONS

1. We are given $A = 35°$, $B = 25°$, $c = 67.6$. Begin by finding C by subtracting:

   $C = 180° - A - B = 180° - 35° - 25° = 120°$.

   Next, by the Law of Sines:

   $$\frac{a}{\sin 35°} = \frac{67.6}{\sin 120°} \text{ or}$$

   $$a = \left(\frac{67.6}{\sin 120°}\right) \times \sin 35° = \left(\frac{67.6}{.8660}\right) \times .5736 = 44.77.$$

   In the same way, we can find b.

   $$\frac{b}{\sin 25°} = \frac{67.6}{\sin 120°} \text{ or}$$

   $$b = \left(\frac{67.6}{\sin 120°}\right) \times \sin 25° = \left(\frac{67.6}{.8660}\right) \times .4226 = 32.99.$$

2. $A = 40°$, $a = 20$, $b = 15$

   Begin by using the Law of Sines to find B:

   $$\sin B = \frac{(b \sin A)}{a} = \frac{15 \sin 40°}{20} = \frac{(15 \times .6428)}{20} = .4821.$$

   Then, we can use the calculator to determine $B = 28.82°$.

   Now, we can find C by subtracting:

   $C = 180° - 40° - 28.82° = 111.18°$.

   Finally, use the Law of Sines to find c:

   $$c = \frac{(a \sin C)}{\sin A} = \frac{(20 \sin 111.18°)}{\sin 40°} = \frac{(20 \times .9324)}{.6428} = 29.01.$$

3. $A = 63°$, $a = 10$, $c = 8.9$.

   Begin by using the Law of Sines to find C:

   $$\sin C = \frac{(c \sin A)}{a} = \frac{8.9 \sin 63°}{10} = \frac{(8.9 \times .8910)}{10} = .7930.$$

   Then, we can use the calculator to determine $C = 52.47°$.

   Now, we can find B by subtracting:

   $B = 180° - 63° - 52.47° = 64.53°$.

   Finally, use the Law of Sines to find b:

   $$b = \frac{(a \sin B)}{\sin A} = \frac{(10 \sin 64.53°)}{\sin 63°} = \frac{(10 \times .9028)}{.8910} = 10.13.$$

SCALARS AND VECTORS

4. $b = 29, c = 17, A = 103°$

   First, we use the Law of Cosines to find a:

   $a^2 = 29^2 + 17^2 - 2(29)(17)\cos 103° = 1351.80$.

   Thus, $a \approx 36.77$.

   Similarly, we can find B from the Law of Cosines:

   $b^2 = a^2 + c^2 - 2ac \cos B$.

   $29^2 = 36.77^2 + 17^2 - 2(36.77)(17)\cos B$.

   Solving for $\cos B$, we get $\cos B = .6398$.

   Use a calculator to determine that $B = 50.22°$.

   Finally, $C = 180° - 50.22° - 103° = 26.78°$.

5. $a = 7, c = 24, c = 26$

   We need to use the Law of Cosines to find the missing angles. We'll find angle A first.

   $a^2 = b^2 + c^2 - 2bc \cos A$

   $7^2 = 24^2 + 26^2 - 2(24)(26)\cos A$

   $\cos A = .9639$. Using a calculator, we determine that $A = 15.43°$.

   Now, use the Law of Cosines in the same way to determine B:

   $b^2 = a^2 + c^2 - 2ac \cos B$

   $24^2 = 7^2 + 26^2 - 2(7)(26)\cos B$

   $\cos B = .4093$.

   Using a calculator, we determine that $B = 65.84°$.

   Finally, $C = 180° - 15.43° - 65.84° = 98.73°$.

6. $A = 18°, B = 47°, C = 115°$

   We do not have enough information to solve this triangle. Recall that if we are given only angles and no sides, there are an infinite number of similar triangle solutions. Therefore, this problem will have no solution.

GRAPHS OF TRIGONOMETRIC FUNCTIONS

In this section, we will take a look at the graphs of the sine, cosine, and tangent functions.

In order to help us draw the graph of $y = \sin \theta$, let's begin by looking at some of the properties of the sine function. First, recall that the function is periodic with period 2π. This means that once we draw the graph for values of θ between 0 and 2π, the graph will simply repeat itself in intervals of 2π. Further, the values for $\sin \theta$ are positive for θ in the first and second quadrants and negative for θ in the third and fourth quadrants. Finally, the values of $\sin \theta$ lie between -1 and $+1$. In fact, since $\sin 0 = 0$, $\sin\left(\dfrac{\pi}{2}\right) = 1$, $\sin \pi = 0$, $\sin\left(\dfrac{3\pi}{2}\right) = -1$, and $\sin 2\pi = 0$, the curve ranges from 0 to 1 to 0 to -1 to 0 and then repeats.

Below is the graph of the sine function. Note how it illustrates the above properties.

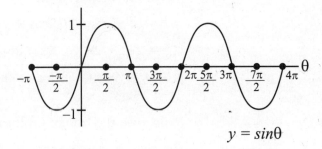

$y = \sin\theta$

The graph of the cosine function has the same shape as that of the sine function. Like the sine function, the cosine function is periodic with period 2π. This tells us that once we draw the graph for values of θ between 0 and 2π, the graph will simply repeat itself in intervals of 2π. Further, the values for $\cos \theta$ are positive for θ in the first and fourth quadrants and negative for θ in the second and third quadrants. Finally, the values of $\cos \theta$ lie between -1 and $+1$. In fact, since $\cos 0 = 1$, $\cos\left(\dfrac{\pi}{2}\right) = 0$, $\cos \pi = -1$, $\cos\left(\dfrac{3\pi}{2}\right) = 0$, and $\cos 2\pi = 1$, the curve ranges from 1 to 0 to -1 to 0 to 1 and then repeats.

The graph of the cosine function is shown below.

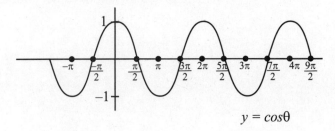

$y = \cos\theta$

GRAPHS OF TRIGONOMETRIC FUNCTIONS

The graph of the tangent function is different in appearance to that of the sine and cosine. Once again, let's examine some of its properties. First, the function is periodic with period π. This tells us that once we draw the graph for values of θ between 0 and π, the graph will simply repeat itself in intervals of π. Next, the values for *tan* θ are positive for θ in the first and third quadrants and negative for θ in the second and fourth quadrants. Finally, the value of *tan* θ can be any real number. In fact, *tan* $0 = 0$, and then the value of the tangent function becomes arbitrarily large as θ approaches $\frac{\pi}{2}$. At $\frac{\pi}{2}$, tangent is undefined. This fact is indicated by the dotted line through $\theta = \frac{\pi}{2}$ in the graph below. On the other side of this dotted line, the tangent function starts out with arbitrarily large negative values, moves up to a value of 0 at π, and then gets bigger again. At $\frac{3\pi}{2}$, tangent is once again undefined.

The graph of the tangent function is given below.

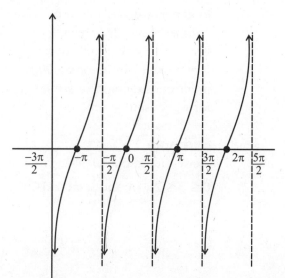

$y = tan\theta$

Next, consider the function $y = 2 \sin \theta$. We can obtain the graph of this function by doubling each y value in the graph $y = \sin \theta$. When we do, we obtain the graph shown below, which is graphed on the same axis as $y = \sin \theta$ for comparison. Notice that the graph has the same shape as the graph of $y = \sin \theta$ but ranges between 2 and -2.

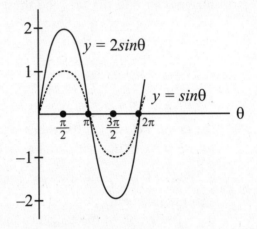

In general, in the graph of either $y = a \sin \theta$ or $y = a \cos \theta$, the number a is called the amplitude and represents the maximum distance of any point on the graph from the x-axis. Thus, the graph of $y = a \sin \theta$ can be obtained by multiplying the y-coordinate of each point on the graph of $\sin \theta$ by the number a. Similar comments apply for the cosine graph.

Example 1

Draw the graphs of $y = \frac{1}{2} \cos \theta$ and $y = 3 \cos \theta$ in the interval $0 \leq \theta \leq 2\pi$ on the same axis.

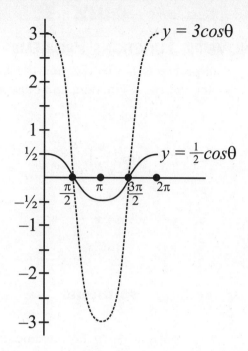

Finally, consider the graph of $y = \sin b\theta$ for $b > 0$. Remember that the graph of $y = \sin \theta$ has a period of 2π. This means that, starting at 0, $y = \sin b\theta$ will repeat its values beginning at $b\theta = 2\pi$, which is to say, when $\theta = \frac{2\pi}{b}$. We say that $y = \sin b\theta$ has a *period* of $\frac{2\pi}{b}$, which means that the graph will repeat itself every $\frac{2\pi}{b}$ units.

Graphs of sine curves and cosine curves with a period different from 2π have the same shape as the regular sine and cosine curve but are "stretched" or "shrunken" in appearance.

Example 2

Sketch the graph of $y = \cos 2\theta$ on the same axis as $y = \cos \theta$.

First of all, note that $y = \cos 2\theta$ has a period of $\frac{2\pi}{2} = \pi$. Thus,

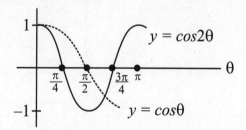

In general, the graph of $y = a \sin b\theta$ or $y = a \cos b\theta$ has amplitude a and period $\frac{2\pi}{b}$.

QUIZ

GRAPHS OF THE TRIGONOMETRIC FUNCTIONS PROBLEMS

In problems 1-5 below, find the amplitude and period of the given sine or cosine function, and draw the graph.

1. $y = 3 \sin 2x$

2. $y = 4 \cos\left(\dfrac{x}{2}\right)$

3. $y = -\cos x$

4. $y = -2 \sin x$

5. $y = -3 \cos(2x)$

SOLUTIONS

1. For $y = 3 \sin 2x$, the amplitude is 3 and the period is $\dfrac{2\pi}{2} = \pi$.

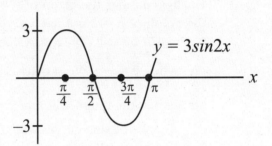

2. For $y = 4 \cos\left(\dfrac{x}{2}\right)$, the amplitude is 4 and the period is $\dfrac{2\pi}{\left(\dfrac{1}{2}\right)} = 4\pi$.

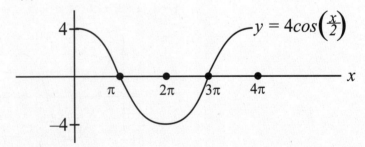

3. For $y = -\cos x$, the amplitude is 1 and the period is 2π. Every y value has the opposite sign of the y values of $y = \cos x$.

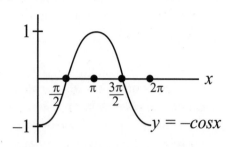

4. For $y = -2 \sin x$, the amplitude is 2 and the period is 2π. Every y value has the opposite sign of the y values of $y = 2 \sin x$.

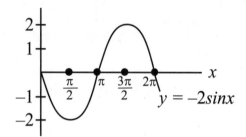

5. For $y = -3 \cos(2x)$, the amplitude is 3 and the period is $\frac{2\pi}{2} = \pi$. Every y value has the opposite sign of the y values of $y = 3 \cos(2x)$.

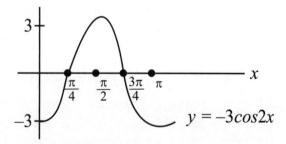

SET THEORY

A knowledge of some basic set theory is required for both the Level IC and Level IIC tests. In addition, set theory will be needed to be able to solve the probability and statistics questions that appear on both tests.

Definitions

A *set* is a collection of objects. The objects in a particular set are called the *members* or the *elements* of the set. In mathematics, sets are usually represented by capital letters, and their members are represented by lower case letters. Braces, { and }, are usually used to enclose the members of a set. Thus, the set A, which has members a, b, c, d, and e and no other members, can be written as A = {a, b, c, d, e}. Note that the order in which the elements of a set are listed is not important; thus, the set {1, 2, 3} and the set {2, 3, 1} represent identical sets.

The symbol used to indicate that an element belongs to a particular set is $\in$, and the symbol that indicates that an element does not belong to a set is $\notin$. Thus, if B = {2, 4, 6, 8}, we can say $6 \in B$ and $7 \notin B$. If a set is defined so that it does not contain any elements, it is called the *empty set*, or the *null set*, and can be written as { } or $\emptyset$.

There are several different notational techniques that can be used to represent a set. The simplest one is called *enumeration*, in which all of the elements of the set are listed within braces. For example, if C is the set of all odd integers between 10 and 20, we can use enumeration to represent the set as C = {11, 13, 15, 17, 19}. The other is called *set-builder notation*. In this notation, a short vertical bar is used to stand for the phrase "such that." For example, the set of all integers less than 15 can be written as:

$\{\, x \mid x < 15, x \text{ is an integer} \,\}$

and is read, *The set of all* x, *such that* x *is less than 15, and* x *is an integer.*

A set that contains a finite number of elements is called a *finite* set. A set that is neither finite nor empty is called an infinite set. When using the method of enumeration to describe a set, we can use three dots to indicate "and so on." Thus, the infinite set containing all positive integers can be written as {1, 2, 3, 4, ...}. The finite set containing all of the even integers between 2 and 200 can be enumerated as {2, 4, 6, ..., 200}.

SET THEORY

Suppose that J is the set containing everyone who lives in New Jersey, and K is the set of all people living in New Jersey who are older that 65. Then, clearly, all members of K are also members of J, and we say *K is a subset of J*. This relationship is written symbolically as $K \subseteq J$. In general, A is a subset of B if every element of A is also an element of B. For example, the set A = {2, 4, 6} is a subset of the set B = {0, 2, 4, 6, 8, 10}. By convention, we agree that the null set is a subset of every other set. Thus, we can write $\emptyset \subseteq A$, where A is any set. Also note that if A and B contain exactly the same elements, then $A \subseteq B$ and $B \subseteq A$. In such a case, we write A = B. If $A \subseteq B$ but $A \neq B$, we call A a *proper subset* of B. This is written $A \subset B$. Thus, if A is a subset of B, and B contains at least one element that is not in A, then A is a proper subset of B, and we write $A \subset B$.

In a particular discussion, the *universal set* represents the largest possible set; that is, it is the set that contains all of the possible elements under consideration. All other sets in the discussion must therefore be subsets of the universal set, which is represented by the letter U. If N is a subset of U, then N′, which is called the *complement* of N, is the set of all elements from the universal set that are not in N. For example, if, in a particular problem, U is the set of all integers, and N is the set of negative integers, then N′ is the set of all non-negative integers.

VENN DIAGRAMS, UNION, AND INTERSECTION

Let U be a universal set and N a subset of U. Then, the drawing below, called a *Venn diagram*, illustrates the relationship between U, N, and N'.

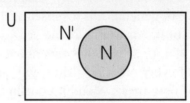

The *union* of two sets A and B, indicated A ∪ B, is the set of all elements that are in either A or B. The intersection of two sets, indicated A ∩ B, is the set of all elements that are in both A and B. Thus, if A = {2, 4, 6, 8, 10} and B = {1, 2, 3, 4}, we have A ∪ B = {1, 2, 3, 4, 6, 8, 10} and A ∩ B = {2, 4}. If A ∩ B = ∅, then A and B are said to be *disjoint*.

The Venn diagrams below represent the operations of union and intersection.

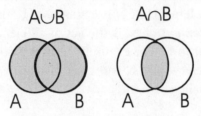

CARTESIAN PRODUCTS

In addition to the operations of union and intersection, there is one other common way of combining two sets. Let A = {1, 2} and B = {3, 4, 5}. Then, the set of all possible ordered pairs (a, b), with a ∈ A and b ∈ B, is called the *Cartesian Product* of A and B and is written A X B. Thus, in this case,

$$A \times B = \{(1,3), (1,4), (1,5), (2,3), (2,4), (2,5)\}$$

QUIZ

Set Problems

1. Use set-builder notation to describe the set of all integers greater than 12 and less than 48.

2. List all of the subsets of the set {a, b, c, d}.

3. If A = {2, 4, 6}, B = {1, 3, 5}, and C = {2, 3, 4}, find A ∪ B, A ∪ C, A ∩ C, A ∩ B, A ∩ (B ∪ C).

4. If U = {2, 4, 6, 8, 10, 12, 14, 16, 18, 20}, and W = {2, 6, 12, 18}, find W'.

5. If Q = {2, 6, 9} and R = {2, 4, 7}, find Q X R.

6. Draw a Venn diagram to represent the set (A ∩ B) ∩ C.

SOLUTIONS

1. {$x \mid 12 < x < 48$, x is an integer}.

2. ∅, {a}, {b}, {c}, {d}, {a, b}, {a, c}, {a, d}, {b, c}, {b, d}, {c, d}, {a, b, c}, {a, b, d}, {a, c, d}, {b, c, d}, {a, b, c, d}.

3. A ∪ B = {1, 2, 3, 4, 5, 6}, A ∪ C = {2, 3, 4, 6}, A ∩ C = {2, 4}, A ∩ B = ∅, A ∩ (B ∪ C) = {2, 4}.

4. W' = {4, 8, 10, 14, 16, 20}.

5. Q X R = {(2,2), (2,4), (2,7), (6,2), (6,4), (6,7), (9,2), (9,4), (9,7)}.

6.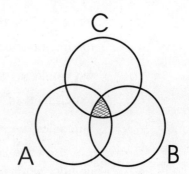

PROBABILITY

Both levels of the SAT II Math test may contain some basic probability questions. In addition, students taking the Level IIC test must know how to count events by using permutations and combinations. Thus, all students should review this section, and those taking the Level IIC test should also study the following section on permutations and combinations.

DEFINITION

Probability is the branch of mathematics that gives you techniques for dealing with uncertainties. Intuitively, probability can be thought of as a numerical measure of the likelihood, or the chance, that an event will occur.

A probability value is always a number between 0 and 1. The nearer a probability value is to 0, the more unlikely the event is to occur; a probability value near 1 indicates that the event is almost certain to occur. Other probability values between 0 and 1 represent varying degrees of likelihood that an event will occur.

In the study of probability, an *experiment* is any process that yields one of a number of well-defined outcomes. By this, we mean that on any single performance of an experiment, one and only one of a number of possible outcomes will occur. Thus, tossing a coin is an experiment with two possible outcomes: heads or tails. Rolling a die is an experiment with six possible outcomes; playing a game of hockey is an experiment with three possible outcomes (win, lose, or tie).

COMPUTING PROBABILITIES

In some experiments, all possible outcomes are equally likely. In such an experiment, with, say, n possible outcomes, we assign a probability of $\frac{1}{n}$ to each outcome. Thus, for example, in the experiment of tossing a fair coin, for which there are two equally likely outcomes, we would say that the probability of each outcome is $\frac{1}{2}$. In the experiment of tossing a fair die, for which there are six equally likely outcomes, we would say that the probability of each outcome is $\frac{1}{6}$.

How would you determine the probability of obtaining an even number when tossing a die? Clearly, there are three distinct ways that an even number can be obtained: tossing a 2, a 4, or a 6. The probability of each one of these three outcomes is $\frac{1}{6}$. The probability of obtaining an even number is simply the sum of the probabilities of these three favorable outcomes; that is to say, the probability of tossing an even number is equal to the probability of tossing a 2, plus the probability of tossing a 4, plus the probability of tossing a 6, which is $\frac{1}{6} + \frac{1}{6} + \frac{1}{6} = \frac{3}{6} = \frac{1}{2}$.

This result leads us to the fundamental formula for computing probabilities for events with equally likely outcomes:

The probability of an event occurring =

$$\frac{\text{The number of favorable outcomes}}{\text{The total number of possible outcomes}}$$

In the case of tossing a die and obtaining an even number, as we saw, there are six possible outcomes, three of which are favorable, leading to a probability of $\frac{3}{6} = \frac{1}{2}$.

Example 1

What is the probability of drawing one card from a standard deck of 52 cards and having it be a king? When you select a card from a deck, there are 52 possible outcomes, 4 of which are favorable. Thus, the probability of drawing a king is $\frac{4}{52} = \frac{1}{13}$.

Example 2

Human eye color is controlled by a single pair of genes, one of which comes from the mother and one of which comes from the father, called a genotype. Brown eye color, B, is dominant over blue eye color ℓ. Therefore, in the genotype Bℓ, which consists of one brown gene B and one blue gene ℓ, the brown gene dominates. A person with a Bℓ genotype will have brown eyes.

If both parents have genotype Bℓ, what is the probability that their child will have blue eyes? To answer the question, we need to consider every possible eye color genotype for the child. They are given in the table below:

father / mother	B	ℓ
B	BB	Bℓ
ℓ	ℓB	$\ell\ell$

The four possible genotypes for the child are equally likely, so we can use the previous formula to compute the probability. Of the four possible outcomes, blue eyes can occur only with the $\ell\ell$ genotype, so only one of the four possible outcomes is favorable to blue eyes. Thus, the probability that the child has blue eyes is $\frac{1}{4}$.

Two events are said to be *independent* if the occurrence of one does not affect the probability of the occurrence of the other. For example, if a coin is tossed and a die is thrown, obtaining heads on the coin and obtaining a 5 on the die are independent events. On the other hand, if a coin is tossed three times, the probability of obtaining heads on the first toss and the probability of obtaining tails on all three tosses are not independent. In particular, if heads is obtained on the first toss, the probability of obtaining three tails becomes 0.

When two events are independent, the probability that they both happen is the product of their individual probabilities. For example, the probability of obtaining heads when a coin is tossed is $\frac{1}{2}$, and the probability of obtaining 5 when a die is thrown is $\frac{1}{6}$; thus, the probability of both of these events happening is

$$\left(\frac{1}{2}\right)\left(\frac{1}{6}\right) = \frac{1}{12}.$$

In a situation where two events occur one after the other, be sure to correctly determine the number of favorable outcomes and the total number of possible outcomes.

Example 3

Consider a standard deck of 52 cards. What is the probability of drawing two kings in a row, if the first card drawn is replaced in the deck before the second card is drawn? What is the probability of drawing two kings in a row if the first card drawn is *not* replaced in the deck?

In the first case, the probability of drawing a king from the deck on the first attempt is $\frac{4}{52} = \frac{1}{13}$. If the selected card is replaced in the deck, the probability of drawing a king on the second draw is also $\frac{1}{13}$, and, thus, the probability of drawing two consecutive kings would be $\left(\frac{1}{13}\right)\left(\frac{1}{13}\right) = \frac{1}{169}$. On the other hand, if the first card drawn is a king and is not replaced, there are now only three kings in a deck of 51 cards, and the probability of drawing the second king becomes $\frac{3}{51} = \frac{1}{17}$. The overall probability, thus, would be $\left(\frac{1}{13}\right)\left(\frac{1}{17}\right) = \frac{1}{221}$.

QUIZ

PROBABILITY PROBLEMS

1. A bag contains 7 blue marbles, three red marbles, and two white marbles. If one marble is chosen at random from the bag, what is the probability that it will be red? What is the probability that it will not be blue?

2. A woman's change purse contains a quarter, two dimes, and two pennies. What is the probability that a coin chosen at random will be worth at least ten cents?

3. A bag contains four white and three black marbles. One marble is selected, its color is noted, and then it is returned to the bag. Then a second marble is selected. What is the probability that both selected marbles were white?

4. Using the same set up as given in problem 3, what is the probability that both selected marbles will be white if the first marble is not returned to the bag?

5. A man applying for his driver's license estimates that his chances of passing the written test are $\frac{2}{3}$, and that his chances of passing the driving test are $\frac{1}{4}$. What is the probability that he passes both tests?

6. If two cards are selected at random from a standard deck of 52 cards, what is the probability that they will both be diamonds?

7. A bag contains 9 marbles, 3 of which are red, 3 of which are blue, and 3 of which are yellow. If three marbles are selected from the bag at random, what is the probability that they are all of different colors?

8. If two standard dice are rolled, what is the probability that the sum of the digits on the two dice is a prime number?

9. What is the probability that if you roll a standard die three times, you will get three different numbers?

10. If you select three cards from a standard deck of 52 cards and they are all kings, what is the probability that the next card you select will also be a king?

SOLUTIONS

1. There are 12 marbles in the bag. Since 3 of them are red, the probability of picking a red marble is $\frac{3}{12} = \frac{1}{4}$. There are 5 marbles in the bag that are not blue, so the probability of picking a marble that is not blue is $\frac{5}{12}$.

2. There are 5 coins in the purse, and 3 of them are worth at least ten cents. Thus, the probability that a coin chosen at random will be worth at least ten cents is $\frac{3}{5}$.

3. There are $7 \times 7 = 49$ ways in which two marbles can be selected. Since there are four ways to select a white marble on the first draw and four ways to select a marble on the second draw, there are a total of $4 \times 4 = 16$ ways to select a white marble on two draws. Thus, the probability of selecting white on both draws is $\frac{16}{49}$.

4. The two selections can be made in $7 \times 6 = 42$ ways. Two white marbles can be selected in $4 \times 3 = 12$ ways. Thus, the desired probability is $\frac{12}{42} = \frac{2}{7}$.

5. Since these two events are independent, the probability of passing both is $\left(\frac{2}{3}\right) \times \frac{1}{4} = \frac{1}{6}$.

6. The probability of drawing a diamond from the full deck is $\frac{13}{52} = \frac{1}{4}$. After the first diamond has been removed, there are 51 cards in the deck, 12 of which are diamonds. The probability of selecting a diamond from this reduced deck is $\frac{12}{51}$. The probability, thus, of selecting two diamonds is $\frac{1}{4} \times \frac{12}{51} = \frac{1}{17}$.

7. After the first marble is selected, the bag has 8 marbles left, 6 of which are of a different color than that of the first marble selected. Thus, the probability that the second marble is of a different color is $\frac{6}{8}$. If the second marble is different, there are then 7 marbles in the bag, three of which are of the color not yet selected. The odds of drawing a marble of the third color is $\frac{3}{7}$. Overall, then, the probability of drawing three different colors is $\frac{6}{8} \times \frac{3}{7} = \frac{18}{56} = \frac{9}{28}$.

8. If two dice are rolled, the possible outcomes for the sums of the two dice are 2 through 12. Of these, 2, 3, 5, 7, and 11 are prime. These is one way to get a sum of two, two ways to get a sum of three, four ways to get a sum of five, six ways to get a sum of seven, and two ways to get a sum of 11. Thus, the probability of rolling a prime sum is $\frac{15}{36} = \frac{5}{12}$.

9. After you roll the die the first time, there is a five out of six chance that the next roll will be different. Then, there is a 4 out of six chance that the third roll will be different. Thus, the probability of rolling three different numbers is $\frac{5}{6} \times \frac{4}{6} = \frac{5}{9}$.

10. After three cards are selected, there are 49 cards left in the deck, of which only one is a king. Thus, the probability of drawing a king on the fourth draw is $\frac{1}{49}$.

PERMUTATIONS AND COMBINATIONS

This section should be studied only by those taking the Level IIC Math test.

A Fundamental Counting Principle

Consider the following problem. A set A contains 3 elements, A = {2, 4, 6}. A set B contains 2 elements, B = {3, 7}. How many different sets exist containing one element from set A and one element from set B?

In order to answer this question, simply note that for each of the three possible selections from set A, there are two possible corresponding selections from set B. Thus, the sets that can be formed are {2, 3}, {2, 7}, {4, 3}, {4, 7}, {6, 3}, and {6, 7}. This means that there are 2 × 3 = 6 sets that can be formed.

This result can be generalized in the following way: If one experiment can be performed in r possible ways, and a second experiment can be performed in s possible ways, then there are a total of rs possible ways to perform both experiments. This principle can be extended to any number of sets and can be applied in many different situations, as the following examples show.

Example 1

How many two-digit numbers can be formed from the digits 2, 4, 6, 8, and 9 if it is permissible to use the same digit twice?

If it is permissible to use the same digit twice, there are 5 choices for the tens digit and 5 choices for the units digit. The principle above, thus, tells us that there are 5 × 5 = 25 ways to form two-digit numbers.

Example 2

How many three-letter sequences can be formed from the letters b, c, d, f, g, and h?

There are six choices for the first letter of the sequence, six choices for the second letter, and six choices for the third. Thus, there are 6 × 6 × 6 = 216 possible sequences.

Permutations

A *permutation* is any arrangement of the elements of a set in definite order. For example, consider the set C = {p, q, r}. There are six different orders in which the elements of this set can be ordered:

pqr, prq, rpq, rqp, qrp, qpr

Thus, there are six permutations of the set C. Of course, it would have been possible to determine that there were six permutations of the given set without listing them all. Simply note that the first letter listed can be any element of the set, so that there are three possible choices for first element. After a letter has been selected to go first, there are two possible selections that remain to go second. And, after the first two selections have been made, there is only one remaining choice for the final selection. By multiplying the number of choices at each stage, $3 \times 2 \times 1 = 6$, we obtain the number of permutations.

The product $3 \times 2 \times 1$ can be written in what is called *factorial notation* as 3!, which is read *three factorial*. Similarly, $5! = 5 \times 4 \times 3 \times 2 \times 1$. And, in general, we have the definition

$$n! = n \times (n-1) \times (n-2) \times \ldots \times 3 \times 2 \times 1$$

The example above illustrates the following fact about permutations: the number of permutations of a set containing n members is $n!$.

Now, consider the following problem. Let D = {a, b, c, d, e, f}. How would we count the number of permutations of the six elements from this set taken three at a time? Once again, we would reason as follows: there are six possible choices for the first element of the permutation, five possible choices for the second element of the permutation, and four choices for the final element. Thus, there are $6 \times 5 \times 4 = 120$ permutations of six objects taken three at a time.

This example illustrates the following fact about permutations: The number of permutations of n elements taken r at a time is given by

$$_nP_r = n(n-1)(n-2) \ldots [n - (r - 1)]$$

Example

In how many ways can a president and a vice president be chosen from a club with eight members?

We are looking for the number of permutations of 8 members taken 2 at a time.

$$_nP_r = n(n - 1) = 8 \times 7 = 56$$

COMBINATIONS

A *combination* is any arrangement of the elements of a set without regard to order. For example, consider the set F = {a, b, c, d}. How many subsets containing three elements does this set have? The subsets are {a, b, c}, {a, b, d}, {a, c, d}, and {b, c, d}. Thus, we say that the number of combinations of four objects taken three at a time, which is written $_4C_3$, is four.

In general, the formula for the number of combinations of n objects taken r at a time is given by

$$_nC_r = \frac{n!}{r!(n-r)!}$$

Example

In how many ways can an advisory board of three members be chosen from a committee of ten?

We need to find the number of combinations of 10 objects taken 3 at a time.

$$_nC_r = \frac{10!}{(3!)(7!)} = \frac{(10 \times 9 \times 8)}{(3 \times 2 \times 1)} = 120$$

QUIZ

PERMUTATIONS AND COMBINATIONS PROBLEMS

1. Brian has five different shirts, two different pairs of pants, and three different ties. How many different outfits can Brian wear?

2. In how many different orders can five boys stand on a line?

3. In how many different ways can a judge award first, second, and third places in a contest with thirteen contestants?

4. A mathematics instructor plans to assign as homework three problems from a set of ten problems. How many different homework assignments are possible?

5. A baseball card dealer has forty different cards that Brian would like to own. For his birthday, Brian is allowed to pick any four of these cards. How many choices does he have?

PERMUTATIONS AND COMBINATIONS

SOLUTIONS

1. By the fundamental counting principle, Brian can wear $5 \times 2 \times 3 = 30$ different outfits.

2. The number of different orders that five boys can stand on a line is given by $5! = 5 \times 4 \times 3 \times 2 \times 1 = 120$.

3. There are thirteen choices for first place. After that, second place can go to twelve people, and then third place can go to eleven people. Thus, there are $13 \times 12 \times 11 = 1,716$ ways to award the prize.

4. Here, we need to count the number of combinations of ten objects taken three at a time. This number is given by

   $$_nC_r = \frac{n!}{r!\,(n-r)!}$$

   with $n = 10$ and $r = 3$. Thus, we need to evaluate:

   $$\frac{n!}{r!(n-r)!} = \frac{10!}{3!(7!)} = \frac{10 \times 9 \times 8}{3 \times 2 \times 1} = 10 \times 3 \times 4 = 120$$

5. We need to count the number of combinations of forty objects taken four at a time. This number is given by

   $$_nC_r = \frac{n!}{r!(n-r)!}$$

   with $n = 40$ and $r = 4$. Thus, we need to evaluate:

   $$\frac{n!}{r!(n-r)!} = \frac{40!}{4!(36!)}$$
   $$= \frac{(40 \times 39 \times 38 \times 37)}{(4 \times 3 \times 2 \times 1)}$$
   $$= 10 \times 13 \times 19 \times 37$$
   $$= 91,390$$

STATISTICS

Statistics is the study of collecting, organizing, and analyzing data. Those taking both levels of the test must know some statistics. Those taking Level I should study the first part of this section: Measures of Location. Those taking Level II should also study the next section: Measures of Variability.

Measures of Location

Measures of location describe the "centering" of a set of data; that is, they are used to represent the central value of the data. There are three common measures of central location. The one that is typically the most useful (and certainly the most common) is the *arithmetic mean*, which is computed by adding up all of the individual data values and dividing by the number of values.

Example 1

A researcher wishes to determine the average (arithmetic mean) amount of time a particular prescription drug remains in the bloodstream of users. She examines five people who have taken the drug and determines the amount of time the drug has remained in each of their bloodstreams. In hours, these times are: 24.3, 24.6, 23.8, 24.0, and 24.3. What is the mean number of hours that the drug remains in the bloodstream of these experimental participants?

To find the mean, we begin by adding up all of the measured values. In this case, $24.3 + 24.6 + 23.8 + 24.0 + 24.3 = 121$. We then divide by the number of participants (five) and obtain $\frac{121}{5} = 24.2$ as the mean.

Example 2

Suppose the participant with the 23.8-hour measurement had actually been measured incorrectly, and a measurement of 11.8 hours obtained instead. What would the mean number of hours have been?

In this case, the sum of the data values is only 109, and the mean becomes 21.8.

This example exhibits the fact that the mean can be greatly thrown off by one incorrect measurement. Similarly, one measurement that is unusually large or unusually small can have great impact upon the mean. A measure of location that is not impacted as much by extreme values is called the *median*. The median of a group of numbers is simply the value in the middle when the data values are arranged in numerical order. This numerical measure is sometimes used in the place of the mean when we wish to minimize the impact of extreme values.

Example 3

What is the median value of the data from Example 1? What is the median value of the modified data from Example 2?

Note that in both cases, the median is 24.3. Clearly, the median was not impacted by the one unusually small observation in Example 2.

In the event that there is an even number of data values, we find the median by computing the number halfway between the two values in the middle (that is, we find the mean of the two middle values).

Another measure of location is called the *mode*. The mode is simply the most frequently occurring value in a series of data. In the examples above, the mode is 24.3. The mode is determined in an experiment when we wish to know which outcome has happened the most often.

Measures of Variability

Measures of location provide only information about the "middle" value. They tell us nothing, however, about the spread or the variability of the data. Yet sometimes knowing the variability of a set of data is very important. To see why, examine the example below.

Consider an individual who has the choice of getting to work using either public transportation or her own car. Obviously, one consideration of interest would be the amount of travel time associated with these two different ways of getting to work. Suppose that over the period of several months, the individual uses both modes of transportation the same number of times and computes the mean for both. It turns out that both methods of transportation average 30 minutes. At first glance, it might appear, therefore, that both alternatives offer the same service. However, let's take a look at the actual data, in minutes:

Travel time using a car: 28, 28, 29, 29, 30, 30, 31, 31, 32, 32

Travel time using public transportation: 24, 25, 26, 27, 28, 29, 30, 33, 36, 42

Even though the average travel time is the same (30 minutes), do the alternatives possess the same degree of reliability? For most people, the variability exhibited for public transportation would be of concern. To protect against arriving late, one would have to allow for 42 minutes of travel time using public transportation, but with a car one would only have to allow a maximum of 32 minutes. Also of concern are the wide extremes that must be expected when using public transportation.

Thus, we can see that when we look at a set of data, we may wish to not only consider the average value of the data but also the variability of the data.

The easiest way to measure the variability of the data is to determine the difference between the largest and the smallest values. This is called the *range*.

Example

Determine the range of the data from Examples 1 and 2.

The range of the data from Example 1 is 24.6 − 23.8 = 0.8. The range of the data in Example 2 is 24.6 − 11.8 = 12.8. Note how the one faulty measurement in Example 2 has changed the range. For this reason, it is usually desirable to use another more reliable measure of variability, called the *standard deviation*.

The standard deviation is an extremely important measure of variability; however, it is rather complicated to compute.

To understand the meaning of the standard deviation, suppose you have a set of data that has a mean of 120 and a standard deviation of 10. As long as this data is "normally distributed" (most reasonable sets of data are), we can conclude that approximately 68 percent of the data values lie within one standard deviation of the mean. This means, in this case, that 68 percent of the data values lie between 120 − 10 = 110 and 120 + 10 = 130. Similarly, about 95 percent of the data values will lie within two standard deviations from the mean; that is, in this case, between 100 and 140. Finally, about 99.7 percent (which is to say, virtually all) of the data will lie within three standard deviations from the mean. In this case, this means that almost all of the data values will fall between 90 and 150.

CORRELATION

Very often, researchers need to determine whether any relationship exists between two variables that they are measuring. For example, they may wish to determine whether an increase in one variable implies that a second variable is likely to have increased as well, or whether an increase in one variable implies that another variable is likely to have decreased.

The *correlation coefficient* is a single number that can be used to measure the degree of the relationship between two variables.

The value of a correlation coefficient can range between −1 and +1. A correlation of +1 indicates a perfect positive correlation; the two variables under consideration increase and decrease together. A correlation of −1 is a perfect negative correlation; when one variable increases, the other decreases, and vice versa. If the correlation is 0, there is no relationship between the behavior of the variables.

Consider a correlation coefficient that is a positive fraction. Such a correlation represents a positive relationship; as one variable increases, the other will tend to increase. The closer that correlation coefficient is to 1, the stronger the relationship will be. Now, consider a correlation coefficient that is a negative fraction. Such a correlation represents a negative relationship; as one variable increases, the other will tend to decrease. The closer the correlation coefficient is to −1, the stronger this inverse relationship will be.

As an example, consider the relationship between height and weight in human beings. Since weight tends to increase as height increases, you might expect that the correlation coefficient for the variables of height and weight would be near +1. On the other hand, consider the relationship between maximum pulse rate and age. In general, maximum pulse rate decreases with age, so you might expect that the correlation coefficient for these two variables would be near −1.

One common mistake in the interpretation of correlation coefficients that you should avoid is the assumption that a high coefficient indicates a cause-and-effect relationship. This is not always the case. An example that is frequently given in statistics classes is the fact that there is a high correlation between gum chewing and crime in the United States. That is to say, as the number of gum chewers went up, there was a similar increase in the number of crimes committed. Obviously, this does not mean that there is any cause and effect between chewing gum and committing a crime. The fact is, simply, that as the population of the United States increased, both gum chewing and crime increased.

The following graphs are three *scatterplots* depicting the relationships between two variables. In the first, the plotted points almost lie on a straight line going up to the right. This is indicative of a strong positive correlation (a correlation near +1). The second scatterplot depicts a strong negative correlation, and the final scatterplot depicts two variables that are unrelated and probably have a correlation that is close to 0.

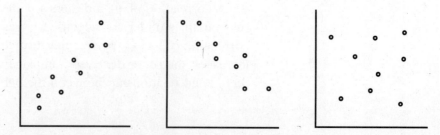

STATISTICS

QUIZ

STATISTICS PROBLEMS

Students taking Level IIC should try to answer all questions; those taking Level IC can answer only questions 1-7.

1. During the twelve months of 1998, an executive charged 4, 1, 5, 6, 3, 5, 1, 0, 5, 6, 4, and 3 business luncheons at the Wardlaw Club. What was the mean monthly number of luncheons charged by the executive?

2. Brian got grades of 92, 89, and 86 on his first three math tests. What grade must he get on his final test to have an overall average of 90?

3. In order to determine the expected mileage for a particular car, an automobile manufacturer conducts a factory test on five of these cars. The results, in miles per gallon, are 25.3, 23.6, 24.8, 23.0, and 24.3. What is the mean mileage? What is the median mileage?

4. In problem 3 above, suppose the car with the 23.6 miles per gallon had a faulty fuel injection system and obtained a mileage of 12.8 miles per gallon instead. What would have been the mean mileage? What would have been the median mileage?

5. In a recent survey, fifteen people were asked for their favorite automobile color. The results were: red, blue, white, white, black, red, red, blue, gray, blue, black, green, white, black, and red. What was the modal choice?

6. An elevator is designed to carry a maximum weight of 3,000 pounds. Is it overloaded if it carries 17 passengers with a mean weight of 140 pounds?

7. The annual incomes of five families living on Larchmont Road are $32,000, $35,000, $37,500, $39,000, and $320,000. What is the range of the annual incomes?

8. The average length of time required to complete a jury questionnaire is 40 minutes, with a standard deviation of 5 minutes. What is the probability that it will take a prospective juror between 35 and 45 minutes to complete the questionnaire?

9. Using the information in problem 8, what is the probability that it will take a prospective juror between 30 and 50 minutes to complete the questionnaire?

10. The scores on a standardized admissions test are normally distributed with a mean of 500 and a standard deviation of 100. What is the probability that a randomly selected student will score between 400 and 600 on the test?

Peterson's: www.petersons.com

SOLUTIONS

1. The mean number of luncheons charged was
$$\frac{(4+1+5+6+3+5+1+0+5+6+4+3)}{12} = \frac{43}{12} = 3.58.$$

2. Let G = the grade on the final test. Then,
$$\frac{(92+89+86+G)}{4} = 90. \quad \text{Multiply by 4.}$$
$(92 + 89 + 86 + G) = 360$
$267 + G = 360$
$G = 93.$
Brian must get a 93 on the final test.

3. The mean mileage is $\frac{(25.3 + 23.6 + 24.8 + 23.0 + 24.3)}{5} = \frac{121}{5}$
= 24.2 miles per gallon. The median mileage is 24.3 miles per gallon.

4. The mean mileage would have been
$\frac{(25.3 + 12.8 + 24.8 + 23.0 + 24.3)}{5} = \frac{110.2}{5} = 22.04$ miles per gallon. The median mileage would have been 24.3 miles per gallon, which is the same as it was in problem 3.

5. The modal choice is red, which was chosen by four people.

6. Since the mean is the total of the data divided by the number of pieces of data, that is mean = $\frac{\text{total}}{\text{number}}$, we have (mean)(number) = total. Thus, the weight of the people on the elevator totals (17)(140) = 2,380. It is therefore not overloaded.

7. The range is $320,000 − $32,000 = $288,000. It can be seen that the range is not a particularly good measure of variability, since four of the five values are within $7,000 of each other.

8. About 68%

9. About 95%

10. About 68%

EXPONENTS AND LOGARITHMS

This section contains topics from advanced algebra. As such, these topics in this section are of importance to those taking the Level II test. Students taking the Level I test can move on to the next section.

DEFINITIONS AND PROPERTIES

When we discussed exponents in the arithmetic review section, all of the exponents we looked at were positive integers. However, meaning can also be given to negative and fractional exponents.

Negative exponents are defined by the definition

$$a^{-n} = \frac{1}{a^n}.$$

Thus, for example, $5^{-2} = \frac{1}{5^2}$, and $7^{-9} = \frac{1}{7^9}$.

By definition, we say that $a^0 = 1$. In other words, any number raised to the 0 power is defined to equal 1. We have therefore given meaning to all integral exponents.

There are five rules for computing with exponents. In general, if k and m are integers, and a and b are any numbers:

Rule 1: $a^k \times a^m = a^{k+m}$

Rule 2: $\dfrac{a^k}{a^m} = a^{k-m}$

Rule 3: $(a^k)^m = a^{km}$

Rule 4: $(ab)^m = a^m \times b^m$

Rule 5: $\left(\dfrac{a}{b}\right)^m = \dfrac{a^m}{b^m}$

Examples

Rule 1: $2^2 \times 2^3 = 4 \times 8 = 32$ and $2^2 \times 2^3 = 2^5 = 32$.

Rule 2: $\dfrac{3^5}{3^7} = \dfrac{243}{2{,}187} = \dfrac{1}{9} = \dfrac{1}{3^2} = 3^{-2}$ and $\dfrac{3^5}{3^7} = 3^{(5-7)} = 3^{-2}$.

Rule 3: $(3^2)^3 = 9^3 = 729$ and $(3^2)^3 = 3^6 = 729$.

Rule 4: $(3 \times 4)^2 = 12^2 = 144$
and $(3 \times 4)^2 = 3^2 \times 4^2 = 9 \times 16 = 144$.

Rule 5: $\left(\dfrac{6}{2}\right)^4 = 3^4 = 81$

and $\left(\dfrac{6}{2}\right)^4 = \dfrac{6^4}{2^4} = \dfrac{1{,}296}{16} = 81$.

SCIENTIFIC NOTATION

Any positive number can be written as the product of a number between 1 and 10 and some power of 10. A number written this way is said to be written in *scientific notation*.

To express a number in scientific notation, begin by repositioning the decimal point so that the number becomes a number between 1 and 10. In other words, place the decimal point so that there is one digit to its left. Then, the appropriate power of 10 can be determined by counting the number of places that the decimal point has been moved. The examples below will clarify this concept.

Example 1

Write the following numbers in scientific notation:

(A) 640,000

In writing this number as 6.4, the decimal point is moved five places to the left. Thus, $640,000 = 6.4 \times 10^5$.

(B) 2,730,000

To change this number to 2.730, the decimal point needs to be moved six places to the left. Thus, $2,730,000 = 2.73 \times 10^6$.

(C) .00085

To change this number to 8.5, the decimal point must be moved four places to the right. Thus, $.00085 = 8.5 \times 10^{-4}$.

(D) .000000562

To change this number to 5.62, the decimal point needs to be moved seven places to the right. Thus, $.000000562 = 5.62 \times 10^{-7}$.

Example 2

Write the following numbers without scientific notation:

(A) 3.69×10^3

Since $10^3 = 1,000$,
we see that $3.69 \times 10^3 = 3.69 \times 1,000 = 3,690$.

(B) 6.7×10^{-4}

Since $10^{-4} = .0001$, $6.7 \times 10^{-4} = 6.7 \times .0001 = .00067$.

EXPONENTS AND LOGARITHMS

FRACTIONAL EXPONENTS

The definitions of exponents can be extended to include fractional exponents. In particular, roots of numbers can be indicated by fractions with a numerator of 1. For example, $\sqrt{2}$ can be written as $2^{\frac{1}{2}}$. Similarly, $\sqrt[3]{7} = 7^{\frac{1}{3}}$. Using rules 1-5 above, we can also make sense of any negative fractional exponents.

Examples

(A) $8^{\frac{-1}{2}} = \frac{1}{\sqrt{8}}$.

(B) $7^{\frac{-5}{2}} = (7^{-5})^{\frac{1}{2}} = \left(\frac{1}{7^5}\right)^{\frac{1}{2}} = \left(\frac{1}{16,807}\right)^{\frac{1}{2}} = \frac{1}{\sqrt{16,807}} \approx .0077$

Note that from Rule 4, we can determine that $(a \times b)^{\frac{1}{2}} = a^{\frac{1}{2}} \times b^{\frac{1}{2}}$. Written in radical notation, this expression becomes $\sqrt{a \times b} = \sqrt{a} \times \sqrt{b}$. This statement justifies the technique we have used for the simplification of square roots.

EXPONENTIAL EQUATIONS

An exponential equation is an equation whose variable appears in a exponent. Such equations can be solved by algebraic means if it is possible to express both sides of the equation as powers of the same base.

Example 1

Solve $5^{2x-1} = 25$.

Rewrite the equation as $5^{2x-1} = 5^2$. Then it must be true that $2x - 1 = 2$. This means that $x = \frac{3}{2}$.

Example 2

Solve $9^{x+3} = 27^{2x}$.

Rewrite the left side of the equation as $(3^2)^{x+3} = 3^{2x+6}$. Rewrite the right side of the equation as $(3^3)^{2x} = 3^{6x}$. Then, it must be true that $2x + 6 = 6x$. This means that $x = \frac{3}{2}$.

Exponential equations in which the bases cannot both be changed to the same number can be solved by using logarithms.

Peterson's: www.petersons.com

THE MEANING OF LOGARITHMS

The logarithm of a number is the power to which a given base must be raised to produce the number. For example, the logarithm of 25 to the base 5 is 2, since 5 must be raised to the second power to produce the number 25. The statement "the logarithm of 25 to the base 5 is 2" is written as $\log_5 25 = 2$.

Note that every time we write a statement about exponents, we can write an equivalent statement about logarithms. For example, $\log_3 27 = 3$ since $3^3 = 27$, and $\log_8 4 = \frac{2}{3}$, since $8^{\frac{2}{3}} = 4$.

An important by-product of the definition of logarithms is that we cannot determine values for $\log_a x$ if x is either zero or a negative number. For example, if $\log_2 0 = b$, then $2^b = 0$, but there is no exponent satisfying this property. Similarly, if $\log_2(-8) = b$, then $2^b = -8$, and there is no exponent satisfying this property.

While logarithms can be written to any base, logarithms to the base 10 are used so frequently that they are called common logarithms, and the symbol "log" is used to stand for "$\log_{10}$."

Examples

1. Write logarithmic equivalents to the following statements about exponents:

    (A) $2^5 = 32$

    The statement $2^5 = 32$ is equivalent to $\log_2 32 = 5$.

    (B) $12^0 = 1$

    The statement $12^0 = 1$ is equivalent to $\log_{12} 1 = 0$.

2. Use the definition of logarithm to evaluate the following:

    (A) $\log_6 36$

    $\log_6 36 = 2$, since $6^2 = 36$.

    (B) $\log_4\left(\frac{1}{16}\right) = -2$, since $4^{-2} = \frac{1}{16}$.

EXPONENTS AND LOGARITHMS

Properties of Logarithms

Since logarithms are exponents, they follow the rules of exponents previously discussed. For example, when exponents to the same base are multiplied and their exponents are added, we have the rule: $\log_a xy = \log_a x + \log_a y$. The three most frequently used rules of logarithms are:

Rule 1: $\log_a xy = \log_a x + \log_a y$

Rule 2: $\log_a \left(\dfrac{x}{y}\right) = \log_a x - \log_a y$

Rule 3: $\log_a x^b = b \log_a x$

Examples

Rule 1: $\log_3 14 = \log_3(7 \cdot 2) = \log_3 7 + \log_3 2$

Rule 2: $\log\left(\dfrac{13}{4}\right) = \log 13 - \log 4$

Rule 3: $\log_7 \sqrt{5} = \log_7(5^{\frac{1}{2}}) = \left(\dfrac{1}{2}\right)\log_7 5$

By combining these rules, we can see, for example, that $\log\left(\dfrac{5b}{7}\right) = \log 5 + \log b - \log 7$.

Solving Exponential Equations by Using Logarithms

Exponential equations, in which neither side can be written as exponents to the same power, can be solved by using logarithms.

Example

Solve $3^{2x} = 4^{x-1}$.

Begin by taking the logarithm of both sides. We could take the logarithm with respect to any base; in this example, to keep things simple, we will take the logarithm to the base 10.

$$\log 3^{2x} = \log 4^{x-1}$$
$$2x \log 3 = (x - 1)\log 4$$
$$2x \log 3 = x \log 4 - \log 4$$
$$2x \log 3 - x \log 4 = -\log 4$$
$$x(2\log 3 - \log 4) = -\log 4$$
$$x = \dfrac{-\log 4}{(2\log 3 - \log 4)}$$

We now need to obtain values for log 3 and log 4. These can be obtained from either a table of logarithms or a scientific calculator. We obtain log 3 = 0.4771 and log 4 = 0.6021. Thus,

$$x = -\frac{(0.6021)}{((2(0.4771)) - 0.6021)} = \frac{-0.6021}{0.3521} = -1.7100$$

QUIZ

EXPONENTS AND LOGARITHMS PROBLEMS

In exercises 1 and 2, write an equivalent exponential form for each radical expression.

1. $\sqrt{11}$
2. $\sqrt[3]{13}$

In exercises 3 and 4, write an equivalent radical expression for each exponential expression.

3. $8^{\frac{1}{5}}$
4. $(x^2)^{\frac{1}{3}}$

In exercises 5 and 6, evaluate the given expressions.

5. $27^{\frac{1}{3}}$
6. $125^{\frac{2}{3}}$

7. Express the following numbers using scientific notation:
   (A) 1,234.56
   (B) 0.0876

8. Write the following numbers without scientific notation:
   (A) 1.234×10^5
   (B) 5.45×10^{-3}

9. Express the following equations in logarithmic form:
   (A) $3^2 = 9$
   (B) $7^{-2} = \frac{1}{49}$

10. Express the following equations in exponential form:
    (A) $\log_6 36 = 2$
    (B) $\log_{10}\left(\frac{1}{10}\right) = -1$

11. Find the value of the following logarithms:
    (A) $\log_2 8$
    (B) $\log_{12} 1$

EXPONENTS AND LOGARITHMS

12. Express as the sum or difference of logarithms of simpler quantities:
    (A) $\log 12$
    (B) $\log\left(\dfrac{ab}{c}\right)$

13. Solve the following equation for x: $125 = 5^{2x-1}$.

SOLUTIONS

1. $11^{\frac{1}{2}}$

2. $13^{\frac{1}{3}}$

3. $\sqrt[5]{8}$

4. $\sqrt[3]{x^2}$

5. $27^{\frac{1}{3}} = \sqrt[3]{27} = 3$

6. $125^{\frac{2}{3}} = (\sqrt[3]{125})^2 = 5^2 = 25$

7. (A) $1{,}234.56 = 1.23456 \times 10^3$
   (B) $0.0876 = 8.76 \times 10^{-2}$

8. (A) $1.234 \times 10^5 = 123{,}400$
   (B) $5.45 \times 10^{-3} = 0.00545$

9. (A) $3^2 = 9$ is equivalent to $\log_3 9 = 2$.
   (B) $7^{-2} = \dfrac{1}{49}$ is equivalent to $\log_7\left(\dfrac{1}{49}\right) = -2$

10. (A) $\log_6 36 = 2$ is equivalent to $6^2 = 36$
    (B) $\log_{10}\left(\dfrac{1}{10}\right) = -1$ is equivalent to $10^{-1} = \dfrac{1}{10}$

11. (A) $\log_2 8 = 3$ (The power that 2 must be raised to in order to equal 8 is 3.)
    (B) $\log_{12} 1 = 0$ (The power that 12 must be raised to in order to equal 1 is 0.)

12. (A) $\log 12 = \log(2^2 \times 3) = \log(2^2) + \log 3 = 2\log 2 + \log 3$
    (B) $\log\left(\dfrac{ab}{c}\right) = \log(ab) - \log c = \log a + \log b - \log c$

13. $125 = 5^{2x-1}$. Rewrite 125 as 5^3. Then,
    $5^3 = 5^{2x-1}$. Thus, it must be true that
    $3 = 2x - 1$ or
    $2x = 4$, so that
    $x = 2$.

LOGIC

This topic should be reviewed by all test-takers. Logic is a field of mathematics in which algebraic techniques are used to establish the truth or falsity of statements.

A *statement* or *assertion* is any expression that can be labeled as either true or false. Letters such as p, q, r, s, and t are used to represent statements. *Compound* statements can be formed by connecting two or more statements.

Examples

1. "My dog is a Boston terrier," "My house is a mess," and "My name is Howard," are statements since they are either true or false.

2. "How old are you?" and "Where do you live?" are not statements since they are neither true nor false.

3. "My cat is named Nora, and my dog is named Krauser," is a compound statement consisting of the two *substatements*, "My cat is named Nora," and "My dog is named Krauser."

4. "He is a natural musician, or he practices a lot," is a compound statement consisting of the two *substatements*, "He is a natural musician," and "He practices a lot."

The truth or falsity of a statement is called its *truth value* (or its logical value). The truth or falsity of a compound statement can be found by determining the truth value of the statements that comprise it and then examining the way in which the statements are connected. We will now consider some of the fundamental ways to connect statements.

CONNECTIVES

If two statements are combined by the word *and*, the resulting compound statement is called the *conjunction* of the two original statements. If p and q are statements, the conjunction of p and q is written $p \wedge q$. In order to determine whether a particular conjunction is true, we can create what is called a *truth table*.

The truth table for conjunction is:

p	q	$p \wedge q$
T	T	T
T	F	F
F	T	F
F	F	F

This first line of this table tells us, for example, that if statement p is true and statement q is true, then $p \wedge q$ is true. On the other hand, the second line tells us that if p is true but q is false, then $p \wedge q$ is false. Note that the table tells us that the only way $p \wedge q$ is true is if both p and q are true.

Example 1

Consider the following four compound statements:

1. The capital of the United States is Washington, D.C., and Christmas is on December 25.

2. The capital of the United States is Washington, D.C., and Christmas is on January 25.

3. The capital of the United States is Buffalo, NY, and Christmas is on December 25.

4. The capital of the United States is Buffalo, NY, and Christmas is on January 25.

Of these four statements, only the first one, which consists of two true substatements, is true.

If two statements are combined by the word *or*, the resulting compound statement is called the *disjunction* of the two original statements. If p and q are two statements, we indicate the disjunction of p and q with the symbol $p \vee q$.

The truth table for disjunction is:

p	q	p ∨ q
T	T	T
T	F	T
F	T	T
F	F	F

Thus, a disjunction is false only when both substatements are false. Note, therefore, that, in logic, the word *or* is used in the sense of *either* or *both*.

Example 2

1. The capital of the United States is Washington, D.C., or Christmas is on December 25.

2. The capital of the United States is Washington, D.C., or Christmas is on January 25.

3. The capital of the United States is Buffalo, NY, or Christmas is on December 25.

4. The capital of the United States is Buffalo, NY, or Christmas is on January 25.

Of these four statements, only the last one, which consists of two false substatements, is false.

Given any statement p, another statement can be created by inserting the word *not* in p. Such a statement is called the *negation* or contradiction of p, is symbolized as $\sim p$, and is read *not p*.

The truth table for negation is:

p	∼p
T	F
F	T

Thus, if p is true, then $\sim p$ is false and vice versa.

For statements p and q, $p \rightarrow q$ represents the statement *if p, then q*. Such statements are called *implications* or *conditional* statements.

LOGIC

The truth table for → is:

p	q	p → q
T	T	T
T	F	F
F	T	T
F	F	T

The statement $p \rightarrow q$ is also read *p implies q*. Another compound statement is the *biconditional* statement, which is written $p \Leftrightarrow q$ and read *p if and only if q*. $p \Leftrightarrow q$ is true if and only if p and q have the same truth values. Thus, the truth table is:

p	q	p ⇔ q
T	T	T
T	F	F
F	T	F
F	F	T

TRUTH TABLES, TAUTOLOGIES, AND CONTRADICTIONS

By using various combinations of the connectives $\wedge$, $\vee$, $\sim$, $\rightarrow$, and $\Leftrightarrow$, we can create compound statements that are much more complicated than those we have considered so far. The truth or falsity of a compound statement can be determined from the truth values of its substatements. A truth table is a simple way to determine the truth or falsity of a compound statement. For example, consider the compound statement $p \vee (q \wedge \sim q)$. A truth table for this statement would look like:

p	q	~q	q ∧ ~q	p ∨ (q ∧ ~q)
T	T	F	F	T
T	F	T	F	T
F	T	F	F	F
F	F	T	F	F

Note that in the table above, the first two columns contain all possible combinations of truth and falsity for p and q. The third and fourth columns simply help us keep track of the truth or falsity of the components of the compound statement, and the final column lists the truth values for the given compound statement. Note that, for example, if p and q are both true, then $p \vee (q \wedge \sim q)$ is also true.

Any compound statement that is true for any possible combination of truth values of the statements that form it is called a *tautology*. In the same way, a compound statement is called a *contradiction* if it is false for all possible truth values of the statements that form it.

Example

Construct a truth table to verify that $p \vee \sim (p \wedge q)$ is a tautology.

p	q	$p \wedge q$	$\sim(p \wedge q)$	$p \vee \sim(p \wedge q)$
T	T	T	F	T
T	F	F	T	T
F	T	F	T	T
F	F	F	T	T

Since $p \vee \sim (p \wedge q)$ is true regardless of the truth or falsity of p and q, it is a tautology.

Two statements are said to be logically equivalent if they have the same truth table. For example, look once again at the truth table for $p \vee (q \wedge \sim q)$. Notice that it has the same values as p. Thus, p and $p \vee (q \wedge \sim q)$ are logically equivalent.

ARGUMENTS, HYPOTHESES, AND CONCLUSIONS

An *argument* is a claim that a number of statements, called *hypotheses* or *premises,* imply another statement, called the *conclusion*. An argument is said to be valid if the conclusion is true whenever all of the hypotheses are true. A *fallacy* is an argument that is not valid. If the hypotheses are written as $p_1, p_2, ..., p_n$, and the conclusion is q, then the arguments are written as $p_1, p_2, ..., p_n \perp q$.

Example

Prove that the argument $p \Leftrightarrow q, q \perp p$ is valid.

Begin by making a truth table containing p, q, and $p \Leftrightarrow q$.

p	q	$p \Leftrightarrow q$
T	T	T
T	F	F
F	T	F
F	F	T

Note that q and $p \Leftrightarrow q$ are both true only on the first line of the table. On this line, p is also true. Thus, whenever q and $p \Leftrightarrow q$ are both true, p is true also. This means that the argument is valid.

Converse, Inverse, and Contrapositive

Once again, let us consider the conditional statement $p \to q$. The statement $q \to p$ is called the *converse* of $p \to q$. Similarly, $\sim p \to \sim q$ is called the *inverse* of $p \to q$, and $\sim q \to \sim p$ is called the *contrapositive* of $p \to q$.

Example

Show that the contrapositive $\sim q \to \sim p$ is logically equivalent to the conditional $p \to q$.

Create the truth tables for the conditional and contrapositive:

p	q	~p	~q	p → q	~q → ~p
T	T	F	F	T	T
T	F	F	T	F	F
F	T	T	F	T	T
F	F	T	T	T	T

Note that the truth tables for the conditional and contrapositive are the same. Therefore, they are logically equivalent.

QUIZ

Logic Problems

1. Let p be the statement, "I am at work," and let q be the statement "It is snowing." Write in words the meaning of each of the following compound statements:

   a. $p \wedge q$
   b. $p \vee q$
   c. $p \wedge \sim q$
   d. $\sim \sim q$
   e. $q \to p$

2. Let r be the statement, "I have a cold," and let s be the statement, "I am at home." Write each of the following using the connective notation discussed in this section.

   a. I have a cold, and I am at home.
   b. I have a cold, and I am not at home.
   c. I do not have a cold.
   d. I do not have a cold, or I have a cold and am at home.
   e. It is false that I do not have a cold and am at home.

3. Construct the truth table for $\sim(\sim p \vee q)$.

4. Demonstrate that $\sim(\sim p \wedge \sim q)$ is logically equivalent to $p \vee q$.

5. Show that "If I go home, then it rained" is logically equivalent to "If it did not rain, then I did not go home."

SOLUTIONS

1. a. I am at work, and it is snowing.
    b. I am at work, or it is snowing.
    c. I am at work, and it is not snowing.
    d. It is false that it is not snowing.
    e. If it is snowing, then I am at work.

2. a. $r \wedge s$
    b. $r \wedge \sim s$
    c. $\sim r$
    d. $\sim r \vee (r \wedge s)$
    e. $\sim (\sim r \wedge s)$

3.
p	q	~p	~p ∨ q	~(~p ∨ q)
T	T	F	T	F
T	F	F	F	T
F	T	T	T	F
F	F	T	T	F

4. First make truth tables for $\sim (\sim p \wedge \sim q)$ and $p \vee q$:

p	q	p ∨ q	~p	~q	~p ∧ ~q	~(~p ∧ ~q)
T	T	T	F	F	F	T
T	F	T	F	T	F	T
F	T	T	T	F	F	T
F	F	F	T	T	T	F

    Note that the third column is the same as the last column. This establishes that the two statements are logically equivalent.

5. Let p represent the statement, "I go home," and let q represent the statement, "It rained." Then, "If I go home, then it rained," can be represented as $p \rightarrow q$. The statement "If it did not rain, then I did not go home," can be represented as $\sim q \rightarrow \sim p$, which is the contrapositive of the given conditional statement. It was shown in a previous problem that the conditional and the contrapositive are logically equivalent.

SYSTEMS OF NUMBERS

This section contains another series of topics that will come up only on the Level IIC test. Those taking that test should pay particular attention to the topic of complex numbers and the way in which they relate to quadratic equations, as complex numbers may turn up in some algebra problems.

Definitions

Recall that within the real number system, numbers of various kinds can be identified. The numbers that are used for counting {1, 2, 3, 4, , ...} are called the *natural numbers* or *positive integers*. The set of positive integers, together with 0, is called the set of *whole numbers*. The positive integers, together with 0 and the *negative integers* {−1, −2, −3, −4, −5, ...}, make up the set of *integers*.

A real number is said to be a *rational number* if it can be written as the ratio of two integers, where the denominator is not 0. Thus, for example, numbers such as −16, ⅔, −⅚, 0, 25, 12⅝ are rational numbers.

Any real number that cannot be expressed as the ratio of two integers is called an *irrational number*. Numbers such as $\sqrt{3}$, $-\sqrt{5}$, and π are irrational. Finally, the set of rational numbers, together with the set of irrational numbers, is called the set of *real numbers*.

Odd and Even Numbers

Any integer that is divisible by 2 is called an *even* integer. Any integer that is not divisible by 2 is called an *odd* integer. When working with odd and even integers, remember the following properties:

Addition:	Subtraction:	Multiplication:
even + even = even	even − even = even	even × even = even
odd + odd = even	odd − odd = even	odd × odd = odd
odd + even = odd	even − odd = odd	even × odd = even
	odd − even = odd	

There are no set rules for division. For example, an even number divided by an even number may be either odd or even.

Prime and Composite Numbers

A whole number N is *divisible* by a whole number M if, when N is divided by M, there is no remainder (that is, it goes in evenly). As an example, since 28 can be evenly divided by 7, we say that 28 is divisible by 7. Further, 7 is said to be a *factor* or a *divisor* of 28, and 28 is said to be a *multiple* of 7. It should be obvious that each whole number has an infinite number of multiples but only a finite number of divisors. For example, the number 15 has multiples 15, 30, 45, 60, 75, 90, ..., but its only divisors are 1, 3, 5, and 15.

A whole number that has only two divisors, itself and 1, is called a *prime number*. A whole number that has more than two divisors is called a *composite* number. Note that 1 and 0 are considered to be neither prime nor composite numbers. Thus, the first seven prime numbers are 2, 3, 5, 7, 11, 13, and 17. On the other hand, the numbers 4, 6, 8, 9, 10, 12, 14, and 15 are composite.

Every composite number can be written as a product of prime numbers in a unique way. For example, the composite number 60 is equal to $2 \times 2 \times 3 \times 5$. This particular factorization of 60 is called its *prime factorization*.

There are several procedures that can be used to prime factor a composite number. Perhaps the easiest technique is simply to make a "factor tree." To begin, write down the number you wish to prime factor. Then, find *any* pair of numbers whose product is the given number, and write this pair of numbers at the end of two branches leading down from the number. Continue this process until every number at the end of each branch is a prime number. The prime factorization is the product of all of the numbers at the ends of the branches.

This process is shown below for the number 315:

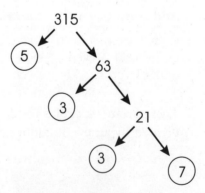

Thus, $315 = 3 \times 3 \times 5 \times 7$

Absolute Value

The *absolute value* of a number is the value of the number without regard to its sign. The absolute value of a number is indicated by placing vertical lines on either side of the number. Thus, for example, |5| = 5, and |–7| = 7.

Example
Calculate the value of |–3| – |8| + |–5| |–2|

|–3| – |8| + |–5| |–2| = 3 – 8 + (5)(2) = 3 – 8 + 10 = 5

The Binary Number System

The number system that we use to represent the numbers we write is called a *decimal* system, since it represents numbers in terms of powers of 10. For example, $1,987 = 1(10)^3 + 9(10)^2 + 8(10) + 7$.

Thus, 1,987 represents a number containing 7 "ones," 8 "tens," 9 "hundreds," and 1 "thousand."

Modern computers, instead, make use of the *binary system*. In this system, numbers are expressed in terms of powers of 2, and only two digits, usually 0 and 1, are required. In this system, the rightmost digit of a number represents the number of "ones" that the number contains, the next digit represents the number of "twos" that the number contains, the next digit represents the number of "fours," then "eights," and so on.

For example, the decimal system number 13 can be written in terms of powers of 2 as: $13 = 1(2)^3 + 1(2)^2 + 0(2) + 1$ and is therefore written in the binary system as $1,101_2$, where the subscript 2 denotes the use of binary digits. The leftmost digit of the figure $1,101_2$ can be thought of as telling us that the decimal number 13 contains one 8. The next digit of $1,101_2$ tells us that the figure contains one 4. Similarly, it contains no 2s and has a remainder of 1.

To write a decimal number in binary, begin by finding the largest power of 2 that is less than the number. Then, subtract this power of 2 from the original number, and find the largest power of 2 that is less than the remainder. Continue this process until you end up with a remainder of 1 or 0.

For example, to express the decimal number 27 in binary, begin by determining the largest power of 2 less than 27. Since the powers of 2 are 1, 2, 4, 8, 16, 32, 64, . . . , the largest power of 2 less than 27 is 16. Put another way, there is one 16 in the number 27. Now, 27 – 16 = 11, and there is one 8 in 11. Next, look at the number 11 – 8 = 3. There are no 4s in 3, but there is one 2, and a remainder of 1. Thus, 27 can be written in binary as 11011_2. Again, think of $1,1011_2$ as representing a number that consists of one 16, one 8, no 4s, one 2, and a remainder of 1.

It is very easy to express a binary number in decimal notation. Consider, for example, the number $101,110_2$. The leftmost digit of this number occupies the "32's place," and, thus, the number contains one 32. It also contains no 16s, one 8, one 4, one 2, and no ones. Thus, $101,110_2 = 32 + 8 + 4 + 2 = 46$.

QUIZ

SYSTEMS OF NUMBERS PROBLEMS

1. If k represents an integer, which of the following must be even integers?

   $2k, 2k + 6, \dfrac{8k}{4}, 3k + 5, 2k - 1$

2. If n represents an odd integer, which of the following must be even integers?

   $2n, 3n, 3n + 1, 3n - 5, 4n + 1$

3. What is the value of $|5| - |-7| - |-4| |12|$?

4. a. Express the number 47 as a binary number.

   b. Express the binary number $110{,}111_2$ in the decimal system.

5. What is the prime factorization of 120?

SOLUTIONS

1. If k represents an integer, then $2k$ is even, since 2 times any integer is even. Further, $2k + 6$ is also even; $\dfrac{8k}{4} = 2k$, so $\dfrac{8k}{4}$ is even. Next, $3k$ may be even or odd, so $3k + 5$ may be even or odd. Finally, $2k - 1$ is odd, since 1 less than an even number is odd. Thus, of the numbers given, $2k$, $2k + 6$, and $\dfrac{8k}{4}$ must be even.

2. If n represents an odd integer, then $2n$ represents an even number times an odd number and is therefore even; $3n$ is the product of two odds and is, thus, odd; $3n + 1$ is the sum of two odds and is, therefore, even; $3n - 5$ is the difference of odds and is, thus, even; $4n + 1$ is the sum of an even and an odd and is, thus, odd. Overall, then, $2n$, $3n + 1$, and $3n - 5$ are even.

3. $|5| - |-7| - |-4| |12| = 5 - (+7) - (4)(12) = 5 - 7 - 48 = -50$

4. a. 47 contains one 32. $47 - 32 = 15$. 15 contains no 16s but contains one 8. $15 - 8 = 7$. 7 contains one 4. $7 - 4 = 3$. 3 contains one 2, with a remainder of one. Therefore, $47 = 101{,}111_2$.

   b. $110{,}111_2 = 32 + 16 + 4 + 2 + 1 = 55$

5. $2 \times 2 \times 2 \times 3 \times 5$

COMPLEX NUMBERS

DEFINITIONS

Up to this point, the real number system has been sufficient for us to be able to solve all of the algebraic equations that we have seen. However, a simple quadratic equation such as

$$x^2 + 1 = 0$$

has no solution in the real number system. In order to be able to solve such equations, mathematicians have introduced the number i, with the property that

$$i^2 + 1 = 0$$

or

$$i^2 = -1$$

Thus, i would be the solution to $x^2 + 1 = 0$.

Since $i^2 = -1$, we can write $i = \sqrt{-1}$ and say that i is equal to "the square root of -1." The number i is called the *imaginary unit*.

Note that when you raise i to successive powers, you end up with values repeating in cycles of four, in the pattern $i, -1, -i, 1$. That is,

$$i^1 = i$$
$$i^2 = -1$$
$$i^3 = (-1)i = -i$$
$$i^4 = (-1)(-1) = 1$$
$$i^5 = (i^4)i = (1)i = i$$
$$i^6 = (i^4)i^2 = (1)i^2 = -1$$
$$i^7 = (i^4)i^3 = (1)i^3 = -i$$
$$i^8 = (i^4)i^4 = (1)(1) = 1, \text{ etc.}$$

Also note that, for any real number n,

$$\sqrt{-n} = \sqrt{(-1)n} = \sqrt{(-1)}\sqrt{n} = i\sqrt{n}$$

Example

Simplify $\sqrt{-25} + \sqrt{-20}$.

$$\sqrt{-25} + \sqrt{-20} = i\sqrt{25} + i\sqrt{20}$$
$$= 5i + 2i\sqrt{5}$$

A number of the form bi, where b is any real number, is called an *imaginary number*. Any number of the form $a + bi$ is called a *complex number*. For any complex number $a + bi$, the number $a - bi$ is called the *complex conjugate*.

OPERATIONS WITH COMPLEX NUMBERS

The following examples show how we perform the fundamental arithmetic operations on complex numbers.

Example 1

Perform the indicated operations:

(a) $(3 + 2i) + (2 + 7i)$
(b) $(3 + 2i) - (2 + 7i)$
(c) $(3 + 2i)(2 + 7i)$
(d) $\dfrac{3 + 2i}{2 + 7i}$

Solutions

(a) $(3 + 2i) + (2 + 7i) = 3 + 2i + 2 + 7i$
$= (3 + 2) + (2i + 7i)$
$= 5 + 9i$

(b) $(3 + 2i) - (2 + 7i) = 3 + 2i - 2 - 7i$
$= (3 - 2) + (2i - 7i)$
$= 1 - 5i$

(c) $(3 + 2i)(2 + 7i)$. Note that complex numbers are multiplied in exactly the same fashion as binomials. Thus,

$(3 + 2i)(2 + 7i) = 6 + 3(7i) + 2(2i) + (2i)(7i)$
$= 6 + 21i + 4i + 14i^2$
$= 6 + 21i + 4i + 14(-1)$
$= 6 + 21i + 4i - 14$
$= -8 + 25i$

(d) $\dfrac{3 + 2i}{2 + 7i}$. The procedure for dividing complex numbers requires that both the top and the bottom of the quotient be multiplied by the *conjugate* of the number on the bottom.

$\dfrac{3 + 2i}{2 + 7i} = \dfrac{3 + 2i}{2 + 7i} \times \dfrac{2 - 7i}{2 - 7i}$

$= \dfrac{(3 + 2i)(2 - 7i)}{(2 + 7i)(2 - 7i)}$

$= \dfrac{6 - 21i + 4i + 14}{4 + 49}$

$= \dfrac{20 - 17i}{53}$

COMPLEX NUMBERS

Frequently, when solving quadratic equations via the quadratic formula, you will end up with complex solutions.

Example 2

Solve the equation $x^2 + 3x + 5 = 0$.

Using the quadratic formula, $a = 1$, $b = 3$, and $c = 5$.

Solution

$$x = \frac{-b \pm \sqrt{b^2 - 4ac}}{2a}$$

$$= \frac{-3 \pm \sqrt{3^2 - 4(1)(5)}}{2(1)}$$

$$= \frac{-3 \pm \sqrt{9 - 20}}{2}$$

$$= \frac{-3 \pm \sqrt{-11}}{2}$$

$$= \frac{-3 \pm i\sqrt{11}}{2}$$

Thus, the solutions are $\frac{-3 + i\sqrt{11}}{2}$ and $\frac{-3 - i\sqrt{11}}{2}$.

QUIZ

COMPLEX NUMBERS PROBLEMS

1. Simplify the following expressions:
   a. $7i^2$
   b. $-6i^8$
   c. $8i^7$

2. Perform the indicated operations:
   a. $(5 + 3i) + (3 - 2i)$
   b. $(8 - 2i) - (-2 - 6i)$

3. Perform the indicated operations:
   a. $(5 + 3i)(3 - 2i)$
   b. $\dfrac{(5 + 3i)}{(3 - 2i)}$

4. Find the sum of $5 + 3i$ and its conjugate.

5. Solve for x: $x^2 + x + 4 = 0$.

SOLUTIONS

1. a. $7i^2 = 7(-1) = -7$
    b. $-6i^8 = -6(1) = -6$
    c. $8i^7 = 8(-i) = -8i$

2. a. $(5 + 3i) + (3 - 2i) = 5 + 3i + 3 - 2i$
    $= 8 + i$
    b. $(8 - 2i) - (-2 - 6i) = 8 - 2i + 2 + 6i$
    $= 10 + 4i$

3. a. $(5 + 3i)(3 - 2i) = 15 - 10i + 9i - (2i)(3i)$
    $= 15 - 10i + 9i + 6$
    $= 21 - i$
    b. $\dfrac{5 + 3i}{3 - 2i} = \dfrac{(5 + 3i)(3 + 2i)}{(3 - 2i)(3 + 2i)}$
    $= \dfrac{15 + 10i + 9i - 6}{9 + 4}$
    $= \dfrac{9 + 19i}{13}$

4. The conjugate of $5 + 3i$ is $5 - 3i$. The sum of the two numbers is simply 10.

5. $x^2 + x + 4 = 0$. We must use the quadratic formula to solve this equation.

    $a = 1, b = 1,$ and $c = 4$.

    $x = \dfrac{-b \pm \sqrt{b^2 - 4ac}}{2a}$

    $= \dfrac{-1 \pm \sqrt{1^2 - 4(1)(4)}}{2(1)}$

    $= \dfrac{-1 \pm \sqrt{1 - 16}}{2}$

    $= \dfrac{-1 \pm \sqrt{-15}}{2}$

    $= \dfrac{-1 \pm i\sqrt{15}}{2}$

SEQUENCES

Both levels of the SAT II Math test contain questions on arithmetic and geometric sequences. In addition, the Level IIC test may also contain a question or two on limits.

Definitions

A collection of numbers of the form

$$a_1, a_2, a_3, ..., a_n$$

is called a *finite sequence*. As the subscripts indicate, a_1 is the first term of the sequence, a_2 is the second term of the sequence, and a_n is the last term of the sequence. In general, we say that a_k is the kth term of the sequence.

A sequence that has no last term is called an *infinite sequence*. An infinite sequence is indicated by writing

$$a_1, a_2, a_3, ..., a_k...,$$

where the three dots at the end indicate that there is a number a_n for every positive integer n.

An equation that gives each term of a sequence as a function of one or more of the preceding terms is called a *recursion formula*. When a sequence of numbers is defined by such a formula, we say that the sequence is defined *recursively*.

Example 1

If the first term of a sequence is 3, and each term thereafter is 5 more than the preceding term, find a recursion formula for each term of the sequence, and find the first five terms of the sequence.

We are given that $a_1 = 3$, and that, for each $k \geq 1$

$$a_{k+1} = a_k + 5$$

Using the equation above with $k = 1, 2, 3$, etc., we compute that $a_2 = a_1 + 5 = 3 + 5 = 8$, $a_3 = 8 + 5 = 13$, etc. Thus, the first five terms of the sequence are:

3, 8, 13, 18, 23.

There are two very important types of recursively defined sequences. The first is called an *arithmetic sequence*.

An arithmetic sequence is one defined by the recursion formula

$$a_{k+1} = a_k + d.$$

The constant d is referred to as the *common difference*. It can be shown that the nth term of the arithmetic sequence with common difference d is given by

$$a_n = a_1 + (n-1)d.$$

Example 2

Find the eighth term of the arithmetic sequence whose first three terms are 21, 12, and 3.

First, determine that the common difference is −9; that is, each term is 9 less than the preceding term. Then, with $a_1 = 21$ and $n = 8$, we have

$$a_8 = 21 + (8 - 1)(-9) = 21 - 63 = -42.$$

One other important formula for arithmetic sequences is the formula that will enable you to compute the sum of the first n terms. Typically, the sum of the first n terms of an arithmetic sequence is denoted as S_n, and then, if the common difference is once again d, we have

$$S_n = \frac{n}{2}[2a_1 + (n-1)d] = \frac{n}{2}(a_1 + a_n).$$

Example 3

Find the sum of the first ten terms of an arithmetic sequence if the first term is −9 and the common difference is 6.

$$S_{10} = (10/2)[2(-9) + (10-1)6] = (5)[-18 + (9)(6)]$$
$$= (5)[-18 + 54] = 180.$$

The other important type of sequence is called a *geometric sequence*. A geometric sequence is defined by the recursion formula

$$a_{k+1} = r\, a_k.$$

The constant r is called the *common ratio*. It is easy to see that in a geometric sequence, each term after the first is obtained by multiplying the preceding term by r. That is, if the first term is a_1, the sequence looks like

$$a_1, a_1 r, a_1 r^2, a_1 r^3, \ldots$$

Just as in the case of the arithmetic sequence, there is a formula for the nth term of a geometric sequence and a formula for the sum of the first n terms of a geometric sequence. First of all, it is easy to see that the nth term of a geometric sequence with common ratio $r \neq 0$ is given by

$$a_n = r^{n-1} a_1.$$

Then, if S_n stands for the sum of the first n terms of a geometric sequence with common ratio $r \neq 1$, we have

$$S_n = a_1 \frac{1 - r^n}{1 - r}.$$

SEQUENCES

Example 4

Find the sum of the first five terms of the geometric sequence whose first term is 6 and whose common ratio is $\frac{1}{3}$.

We have

$$S_5 = 6\,\frac{1-\left(\frac{1}{3}\right)^5}{1-\frac{1}{3}}$$

$$= 6\,\frac{1-\left(\frac{1}{3}\right)^5}{1-\frac{1}{3}} \times \frac{3}{3}$$

$$= 6\,\frac{3-\left(\frac{1}{3}\right)^4}{3-1}$$

$$= 3\left[3-\left(\frac{1}{3}\right)^4\right]$$

$$= 9 - \frac{1}{27}$$

$$= 8\frac{26}{27}.$$

LIMITS OF INFINITE SERIES

Consider an infinite sequence. It may initially appear impossible to find the sum of all of the terms of such a sequence, since there are an infinite number of terms. However, sometimes it happens that as k gets larger and larger, the sum of the first k terms, S_k, gets closer and closer to some definite finite value S. Whenever this happens, we define this value S to be the sum of the infinite series and say that S_k *converges to S*. This is sometimes written as $S_k \to S$, or as $\lim_{k \to \infty} S_k = S$. In either case, we say that the limit of S_k as k approaches infinity is S.

Example

Consider the infinite geometric sequence with first term 1 and common ratio $\frac{1}{2}$. Find the sum of this sequence.

Let's begin by finding an expression for the sum of the first n terms of the sequence. Using the formula

$$S_n = a_1 \frac{1 - r^n}{1 - r},$$

we obtain

$$S_n = a_1 \frac{1 - r^n}{1 - r}$$

$$= 1 \frac{1 - \left(\frac{1}{2}\right)^n}{1 - \left(\frac{1}{2}\right)} = \frac{1 - \left(\frac{1}{2}\right)^n}{\frac{1}{2}} = 2 - 2\left(\frac{1}{2}\right)^n = 2 - \left(\frac{1}{2}\right)^{n-1}.$$

Now, clearly, as n gets larger and larger, the second term in this expression gets closer and closer to 0. Therefore, it appears as if the limit of S_n as n approaches infinity is $2 - 0 = 2$.

SEQUENCES

QUIZ

SEQUENCES PROBLEMS

Note: Those taking Level IC should attempt the first eight questions. Those taking Level IIC should answer all 10 questions.

1. Find the first five terms of the sequence specified by the recursion formula $a_{k+1} = a_k + 3$, if $a_1 = 7$.

2. Find the tenth term of the sequence defined in problem 1.

3. Find the sum of the first 10 terms of the sequence from problem 1.

4. Find the first five terms of the sequence specified by the recursion formula $a_{k+1} = 3a_k$, if $a_1 = 2$.

5. Find the tenth term of the sequence defined in problem 4.

6. Find the sum of the first 10 terms of the sequence from problem 4.

7. Find the sum of all of the odd integers between 200 and 400.

8. Find the sum of all of the integers between 2 and 322 that are divisible by 3.

9. Consider the infinite geometric sequence with first term 5 and common ratio $\frac{1}{2}$. Find the sum S of the terms of this sequence, if it exists.

10. Find the first three terms of the infinite geometric sequence with sum 81 and common ratio $\frac{1}{3}$.

MATHEMATICS IC AND IIC REVIEW

SOLUTIONS

1. This arithmetic sequence has first term 7 and common difference 3. Thus, the first five terms are 7, 10, 13, 16, and 19.

2. To find the tenth term of the sequence, use the formula $a_n = a_1 + (n - 1)d$.

   We have $a_{10} = 7 + (9)(3) = 7 + 27 = 34$.

3. Use the formula $S_n = \frac{n}{2}[2a_1 + (n - 1)d] = \frac{n}{2}(a_1 + a_n)$.

   We have $S_{10} = \left(\frac{10}{2}\right)(7 + 34) = (5)(41) = 205$.

4. This sequence is geometric with common ratio 3 and first term 2. Thus, the first five terms are 2, 6, 18, 54, and 162.

5. To find the tenth term of the sequence, use the formula $a_n = r^{n-1}a_1$.

   We obtain $a_{10} = (3^9)(2) = (19{,}683)(2) = 39{,}366$.

6. To find the sum of the first ten terms, use the formula
   $S_n = a_1 \frac{1 - r^n}{1 - r}$. We obtain

   $S_{10} = 2\dfrac{1 - 3^{10}}{1 - 3} = 2\dfrac{59{,}048}{2} = 59{,}048$.

7. The sequence of odd integers between 200 and 400 begins with 201, ends with 399, has a common difference of 2, and contains 100 terms. To find the sum, use the formula
   $S_n = \frac{n}{2}[2a_1 + (n - 1)d] = \frac{n}{2}(a_1 + a_n)$. We obtain

   $S_{100} = \dfrac{100}{2}(201 + 399) = 50(600) = 30{,}000$.

8. This sequence begins with 3, ends with 321, and contains 107 terms. (There are 100 numbers divisible by 3 between 2 and 301 and 7 more terms divisible by 3 between 301 and 322.)

   Using the formula $S_n = \frac{n}{2}[2a_1 + (n - 1)d] = \frac{n}{2}(a_1 + a_n)$, we obtain

   $S_{107} = \dfrac{107}{2}(3 + 321) = \dfrac{107}{2}(324) + 107(162) = 17{,}334$.

SEQUENCES

9. The formula for the sum of a geometric sequence is $S_n = a_1 \dfrac{1 - r^n}{1 - r}$. Substituting the values for a_1 and r, we get

$$S_n = 5 \dfrac{1 - \left(\dfrac{1}{2}\right)^n}{1 - \dfrac{1}{2}} = 5 \dfrac{1 - \left(\dfrac{1}{2}\right)^n}{\dfrac{1}{2}} = 10\left[1 - \left(\dfrac{1}{2}\right)^n\right] = 10 - 10\left(\dfrac{1}{2}\right)^n.$$

Now, as n gets arbitrarily large, $\left(\dfrac{1}{2}\right)^n$ grows arbitrarily close to zero, and, therefore, $10\left(\dfrac{1}{2}\right)^n$ gets arbitrarily close to 0. Therefore, S is equal to 10.

10. In order to solve this problem, we once again will need the formula $S_n = a_1 \dfrac{1 - r^n}{1 - r}$. We are told that the common ratio is $\dfrac{1}{3}$, and that as n gets arbitrarily large, S_n approaches 81. First of all, then,

$$S_n = a_1 \dfrac{1 - r^n}{1 - r} = a_1 \dfrac{1 - \left(\dfrac{1}{3}\right)^n}{1 - \dfrac{1}{3}}$$

$$= a_1 \dfrac{1 - \left(\dfrac{1}{3}\right)^n}{\dfrac{2}{3}}$$

$$= \dfrac{3a_1}{2}\left[1 - \left(\dfrac{1}{3}\right)^n\right]$$

$$= \dfrac{3a_1}{2} - \dfrac{3a_1}{2}\left(\dfrac{1}{3}\right)^n.$$

Now, as n approaches infinity, S_n approaches 81 and the term $\dfrac{3a_1}{2}\left(\dfrac{1}{3}\right)^n$ will approach 0. Overall, then, as n approaches infinity, we obtain $81 = \dfrac{3a_1}{2}$. From this, we see that $a_1 = 54$. The second term is $\dfrac{54}{3} = 18$, and the third term is $\dfrac{18}{3} = 6$.

Practice Test 1

MATHEMATICS LEVEL IC TEST

PRACTICE TEST 1

MATHEMATICS LEVEL IC TEST

While you have taken many standardized tests and know to blacken completely the ovals on the answer sheets and to erase completely any errors, the instructions for the SAT II Mathematics IC exam differs in three important ways from the directions for other standardized tests you have taken. You need to indicate on the answer key which test you are taking.

The instructions on the answer sheet will tell you to fill out the top portion of the answer sheet exactly as shown.

1. Print MATHEMATICS LEVEL IC on the line to the right under the words *Subject Test (print)*.

2. In the shaded box labeled *Test Code* fill in four ovals:

    —Fill in oval 3 in the row labeled V.
    —Fill in oval 2 in the row labeled W.
    —Fill in oval 5 in the row labeled X.
    —Fill in oval A in the row labeled Y.
    —Leave the ovals in row Q blank.

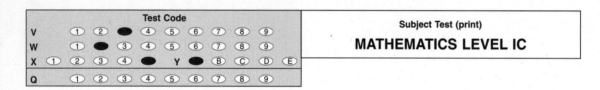

When everyone has completed filling in this portion of the answer sheet, the supervisor will tell you to turn the page and begin the Mathematics Level IC examination. The answer sheet has 100 numbered ovals on the sheet, but there are only 50 multiple-choice questions in the test, so be sure to use only ovals 1 to 50 to record your answers.

MATHEMATICS LEVEL IC TEST

REFERENCE INFORMATION

The following information is for your reference in answering some of the questions in this test.

Volume of a right circular cone with radius r and height h:
$$V = \frac{1}{3}\pi r^2 h$$

Lateral area of a right circular cone with circumference of the base c and slant height l:
$$S = \frac{1}{2}cl$$

Surface area of a sphere with radius r:
$$S = 4\pi r^2$$

Volume of a pyramid with base area B and height h:
$$V = \frac{1}{3}Bh$$

For each of the following problems, identify the BEST answer of the choices given. If the exact numerical value is not one of the choices, select the answer that is closest to this value. Then fill in the corresponding oval on the answer sheet.

NOTES

1. You will need a calculator to answer some (but not all) of the questions. You must decide whether or not to use a calculator for each question. You must use at least a scientific calculator; you are permitted to use graphing or programmable calculators.

2. Degree measure is the only angle measure used on this test. Be sure your calculator is set in degree mode.

3. The figures that accompany questions on the test are designed to give you information that is useful in solving problems. They are drawn as accurately as possible EXCEPT when stated that a figure is not drawn to scale. Unless otherwise indicated, all figures lie in planes.

4. The domain of any function f is assumed to be the set of all real numbers x for which $f(x)$ is a real number, unless otherwise specified.

PRACTICE TEST 1

MATHEMATICS LEVEL IC TEST

USE THIS SPACE FOR SCRATCH WORK

1. If $5x + 3x = 2x + 7x - 5$, $x =$

   (A) -5
   (B) 0
   (C) 5
   (D) $\dfrac{5}{17}$
   (E) $-\dfrac{5}{17}$

2. If $t = 2$, $(t - 4)(t + 3) =$

   (A) -2
   (B) -10
   (C) 5
   (D) 10
   (E) 3

3. For all $x \neq 1$, $\dfrac{3}{\dfrac{2}{(x-1)^2}} =$

   (A) $\dfrac{3(x-1)^2}{2}$
   (B) $\dfrac{6}{(x-1)^2}$
   (C) $\dfrac{(x-1)^2}{6}$
   (D) $\dfrac{2(x-1)^2}{3}$
   (E) $6(x-1)^2$

MATHEMATICS LEVEL IC TEST—*Continued*

USE THIS SPACE FOR SCRATCH WORK

4. $(x + y - 2)(x + y + 2) =$

(A) $x^2 + y^2 + 4$
(B) $(x + y)^2 - 4(x + y) - 4$
(C) $(x + y)^2 - 4(x + y) + 4$
(D) $x^2 + y^2 - 4$
(E) $(x + y)^2 - 4$

5. If $3x^2 = 5$, $(9x^2)^2 =$

(A) 75
(B) 15
(C) 125
(D) 225
(E) 25

6. In the square *ABCD*, each side has a length of 5. The coordinates of vertex *C* are

(A) (6, 1)
(B) (1, 6)
(C) (6, 6)
(D) (5, 5)
(E) (1, 5)

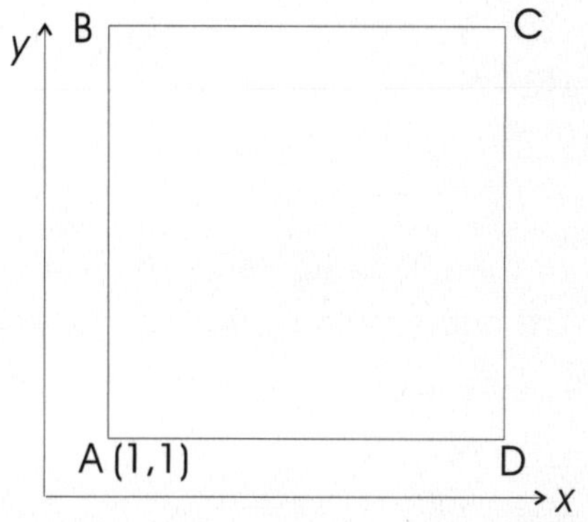

7. The point of intersection of the graph $3x - 2y + 6 = 0$ with the y-axis is

(A) (0, −3)
(B) (0, 3)
(C) (0, 6)
(D) (0, −6)
(E) (0, −2)

GO ON TO THE NEXT PAGE

8. For three bags, A, B, and C, bag A contains one half the number of apples as bag B, and bag B contains one third the number of apples as bag C. If bag A has 30 apples, how many does bag C have?

   (A) 5
   (B) 20
   (C) 180
   (D) 45
   (E) 90

9. The cube of the square root of a number is 27. The number is

   (A) $\sqrt{3}$
   (B) 18
   (C) 81
   (D) 3
   (E) 9

10. In the following figure, lines A and B are parallel and are intersected by line C. The value of angle m is

    (A) 120°
    (B) 60°
    (C) 300°
    (D) 420°
    (E) 240°

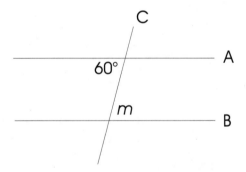

11. If $x + y = 5$ and $x + 2y = 7$, then $x =$

    (A) 3
    (B) 2
    (C) 0
    (D) 5
    (E) 4

12. If $2x + y + 3x = 2y$, $x + y =$

   (A) $5x + y$
   (B) $5x$
   (C) $4x$
   (D) $7x$
   (E) $6x$

13. If $f(x) = 2x - 1$, $f(2.2) =$

   (A) 1
   (B) 3.4
   (C) $2x - 2.2$
   (D) 4.4
   (E) 2.2

14. If $sin\theta = 0.6$, $cos\theta =$

   (A) 0.4
   (B) 0.64
   (C) 0.8
   (D) 0.5
   (E) 1

15. If $2^3 \cdot 2^2 = 2^k$, the value of $k =$

   (A) 5
   (B) 6
   (C) 3
   (D) 2
   (E) 1

MATHEMATICS LEVEL IC TEST—Continued

16. What is the value of $f(2x - 3)$, if $f(x) = \frac{1}{x^2}$ and $x \neq 0$?

    (A) $\frac{1}{2x^2 - 3}$
    (B) $2x - 3$
    (C) $2x^2 - 3$
    (D) $\frac{1}{(2x - 3)^2}$
    (E) $\frac{1}{2x - 3}$

17. If your daily income increases from $60 to $72, what is the percentage of increase?

    (A) 12%
    (B) 16%
    (C) 84%
    (D) 120%
    (E) 20%

18. If $|x - 3| > 2$, which of the following is true?

    (A) $-5 < x < 1$
    (B) $x = 1$
    (C) $x < 1 \text{ or } x > 5$
    (D) $x = 5$
    (E) $1 < x < 5$

19. A rectangular box of size $100 \times 50 \times 20$ cubic cm has to be wrapped. If wrapping paper costs $1.20 per square meter, what will it cost to wrap the box?

    (A) $1.20
    (B) $2.40
    (C) $2.04
    (D) $1.92
    (E) $0.96

20. If 5 peaches sell for $7.50, how much will 8 peaches sell for?

    (A) $1.50
    (B) $16
    (C) $12
    (D) $15
    (E) $10

21. If a mixture of 2 compounds, A and B, contains 30% A and 70% B, how many mg of A are present in 30 mg of the compound?

    (A) 9 mg
    (B) 21 mg
    (C) 5 mg
    (D) 15 mg
    (E) 25 mg

22. If a line A is parallel to line $y = 3x - 6$ and it cuts the y-axis at point (0, 4), then the equation of the line is

    (A) $y = 3x - 4$
    (B) $y = 4x$
    (C) $y = 3x + 4$
    (D) $y = \dfrac{-x}{3} + 4$
    (E) $y = 3x$

MATHEMATICS LEVEL IC TEST—*Continued*

23. If the polygon to the right is a square with sides that are 6 each, then the length of segment *BD* is

    (A) 72
    (B) 36
    (C) 12
    (D) $6\sqrt{2}$
    (E) 6

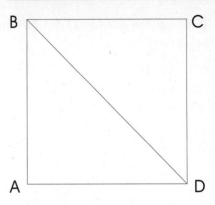

24. In triangle *ABC*, *AB* = *AC* and ∠*ABC* = 55°. What is the value of ∠*BAC*?

    (A) 55°
    (B) 70°
    (C) 110°
    (D) 60°
    (E) 35°

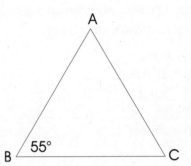

25. A regular hexagon is made up of 6 equilateral triangles. Segment *OA* is 3 cm long. The perimeter of the hexagon is

    (A) 9
    (B) 15
    (C) 18
    (D) 27
    (E) 36

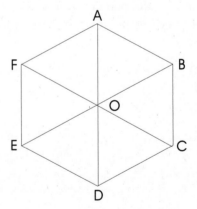

26. A road on a hill is elevated at an angle of 30°. If a man drives 5 meters on the road, how high does he go above the starting point?

    (A) 2.5 m
    (B) 5 m
    (C) 4.3 m
    (D) 1.73 m
    (E) 3.25 m

27. If the midpoint of a horizontal line segment with a length of 8 is (3, −2), then the coordinates of its endpoints are

   (A) (1, −2) and (9, −2)
   (B) (3, −6) and (3, 2)
   (C) (3, 0) and (3, 8)
   (D) (−1, −2) and (7, −2)
   (E) (0, −2) and (8, −2)

28. What is the length of chord AB of the circle centered at the origin in the following figure?

   (A) 8
   (B) 1.414
   (C) 4
   (D) 32
   (E) 5.656

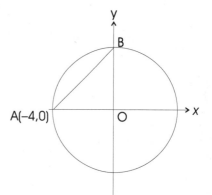

29. In the following figure, the length of arc ACB is

   (A) 31.4
   (B) 0.393
   (C) 3.142
   (D) 3.93
   (E) 5

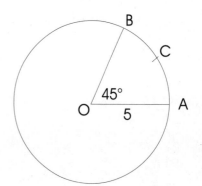

MATHEMATICS LEVEL IC TEST—Continued

USE THIS SPACE FOR SCRATCH WORK

30. The solid in the figure below is the part of a cone with a height of 15 cm obtained by cutting its top parallel to the base, 10 cm from its tip. The volume of the solid is

    (A) 188.52
    (B) 23.57
    (C) 164.93
    (D) 145.29
    (E) 152.68

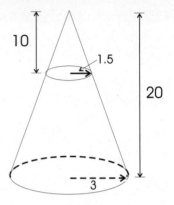

31. The area of an isosceles right triangle inscribed in the circle with center at O and radius $5\sqrt{2}$ is

    (A) 100
    (B) 50
    (C) $10\sqrt{2}$
    (D) $25\sqrt{2}$
    (E) $50\sqrt{2}$

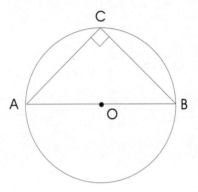

32. $\sqrt{\dfrac{2}{3}}$ can also be written as

    (A) $\dfrac{2}{\sqrt{3}}$
    (B) $\dfrac{\sqrt{2}}{3}$
    (C) $\dfrac{4}{9}$
    (D) $\dfrac{\sqrt{3}}{3}$
    (E) $\dfrac{\sqrt{6}}{3}$

33. A three-character combination lock is made of digits and letters chosen from the following three sets:

    $A = \{0, 1, 2,...9\}$, $B = \{A, B, C,...Z\}$, and $C = \{0, 1, 2,...9\}$.

    If the combination is such that each of the three positions can be one digit or letter from A, B, and C respectively, how many unique combinations are possible?

    (A) 126
    (B) 36
    (C) 2,600
    (D) 520
    (E) 270

34. If an ant completely traverses a circular path in 12 seconds, how many degrees does it turn by in 2 seconds?

    (A) 30°
    (B) 60°
    (C) 15°
    (D) 90°
    (E) 120°

35. In the following figure, what is the value of x if $\theta = 35°$?

    (A) 9.83
    (B) 8.29
    (C) 7.63
    (D) 8.40
    (E) 10.12

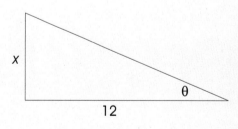

MATHEMATICS LEVEL IC TEST—Continued

36. The acceleration of a car is a function of the time elapsed since it started to accelerate. If t represents the time lapsed since the car started to accelerate and $a(t)$ represents the acceleration of the car, given by $a(t) = 16t^2 + 100t + 75$, what is the acceleration of the car 5 seconds after it started to accelerate?

(A) 900
(B) 975
(C) 1,050
(D) 1,195
(E) 850

37. $(sin\theta + cos\theta)^2$ is the same as

(A) $sin^2\theta + cos^2\theta$
(B) 1
(C) $1 + 2sin\theta \, cos\theta$
(D) $1 - 2 sin\theta \, cos\theta$
(E) $sin^2\theta + 1$

38. In the figure to the right, triangles ABC and CED are similar. What is the value of $\dfrac{x}{y}$?

(A) $\dfrac{3}{7}$
(B) $\dfrac{4}{7}$
(C) $\dfrac{4}{55}$
(D) $\dfrac{3}{4}$
(E) $\dfrac{4}{3}$

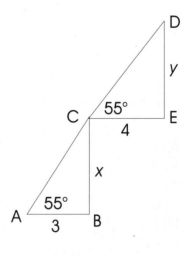

MATHEMATICS LEVEL IC TEST—Continued

USE THIS SPACE FOR SCRATCH WORK

39. For function f, where $f(x) = (1 + 2x)^2$ is defined for $-1 \leq x \leq 2$, what is the range of f?

    (A) $0 \leq f(x) \leq 5$
    (B) $-1 \leq f(x) \leq 2$
    (C) $1 \leq f(x) \leq 25$
    (D) $0 \leq f(x) \leq 25$
    (E) $1 \leq f(x) \leq 5$

40. Town A is located at point (3, 5) on the map and town B is at point (8, 11). What is the shortest distance between the two towns?

    (A) $\sqrt{61}$
    (B) 61
    (C) $\sqrt{11}$
    (D) 11
    (E) $\sqrt{13}$

41. A certain bacteria is growing at a rate of 4 percent each day. If the culture of bacteria has 5,000 bacteria today, how many bacteria will be present 10 days from now?

    (A) 7,500
    (B) 7,400
    (C) 7,300
    (D) 7,200
    (E) 7,100

GO ON TO THE NEXT PAGE

MATHEMATICS LEVEL IC TEST—Continued

USE THIS SPACE FOR SCRATCH WORK

42. In the figure to the right, $p = 70°$ and $90 < q° < 150$. Then, the value of $(r + s)$ can be described as

    (A) $0 \leq r + s \leq 290$
    (B) $290 \leq r + s \leq 360$
    (C) $0 \leq r + s \leq 70$
    (D) $140 \leq r + s \leq 200$
    (E) $90 \leq r + s \leq 150$

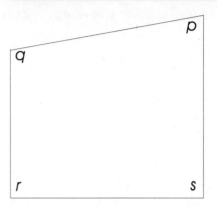

43. Consider n buckets arranged in a row. We put one marble in the first one. Every successive bucket gets twice the number of marbles as the previous bucket. How many marbles will the nth bucket contain?

    (A) $2n$
    (B) $1 + 2(n - 1)$
    (C) $2(n - 1)$
    (D) 2^n
    (E) 2^{n-1}

44. In the right circular cylinder shown to the right, O and X are the centers of the bases and AB is a diameter of one of the bases. If the perimeter of $\triangle AXB$ is 36 and the radius of base of the cylinder is 5, the height of the cylinder is

    (A) 12
    (B) 13.66
    (C) 14.66
    (D) 11.37
    (E) 10.85

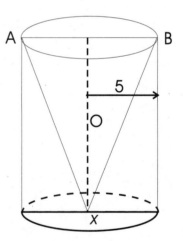

45. If $f(x) = x^2 + 2x - 1$ and $g(t) = t + 1$, $f(g(t)) =$

    (A) $(x^2 + 2x - 1) + (t + 1)$
    (B) $(x^2 + 2x - 1)(t + 1)$
    (C) $t^2 + 4t + 2$
    (D) $(x^2 + 2x - 1) + 1$
    (E) $t^2 + 2t - 1$

MATHEMATICS LEVEL IC TEST—Continued

USE THIS SPACE FOR SCRATCH WORK

46. One ball is drawn from a bag containing 3 white, 4 red, and 5 blue balls. What is the probability that it is a white or red ball?

    (A) $\dfrac{12}{144}$

    (B) $\dfrac{7}{144}$

    (C) $\dfrac{3}{12}$

    (D) $\dfrac{7}{12}$

    (E) $\dfrac{4}{12}$

47. The equations of asymptotes to the hyperbola $\dfrac{x^2}{36} - \dfrac{y^2}{25} = 1$ are

    (A) $y = \dfrac{25}{36}x$ and $y = \dfrac{-25}{36}x$

    (B) $y = \dfrac{5}{6}x$ and $y = \dfrac{-5}{6}x$

    (C) $y = \dfrac{36}{25}x$ and $y = \dfrac{-36}{25}x$

    (D) $y = \dfrac{6}{5}x$ and $y = \dfrac{-6}{5}x$

    (E) $y = \dfrac{1}{25}x$ and $y = \dfrac{1}{36}x$

GO ON TO THE NEXT PAGE

MATHEMATICS LEVEL IC TEST—Continued

USE THIS SPACE FOR SCRATCH WORK

48. In a group of 10 people who have salaries ranging from $75,000 to $200,000, the salary of the person earning the maximum is increased from $200,000 to $225,000. What is true about the current mean and median as compared to before the salary increase?

    (A) The mean and median remain the same.
    (B) The mean increases but the median stays the same.
    (C) The mean stays the same but the median increases.
    (D) Both the mean and the median increase.
    (E) None of the above.

49. What is the value $(sin^2\theta + cos^2\theta - 4)^2$ if $sin\theta = 0.63$?

    (A) 3
    (B) 9
    (C) 6.78
    (D) 10.17
    (E) 8.39

50. $(a^2)^3 + (b^2)(b^3) =$

    (A) $a^5 + b^5$
    (B) $a^6 + b^6$
    (C) $a^5 + b^6$
    (D) $a^6 + b^5$
    (E) $a^6 b^5$

STOP If you finish before time is called, you may check your work on this section only. Do not turn to any other section in the test.

Quick Score Answers

1. C	11. A	21. A	31. B	41. B
2. B	12. E	22. C	32. E	42. D
3. A	13. B	23. D	33. C	43. E
4. E	14. C	24. B	34. B	44. A
5. D	15. A	25. C	35. D	45. C
6. C	16. D	26. A	36. B	46. D
7. B	17. E	27. D	37. C	47. B
8. C	18. C	28. E	38. D	48. B
9. E	19. D	29. D	39. D	49. B
10. B	20. C	30. C	40. A	50. D

ANSWERS AND EXPLANATIONS

1. **The correct answer is (C).** $5x + 3x = 2x + 7x - 5$
$$5x + 3x - 2x - 7x = -5$$
$$-x = 5$$
$$x = 5$$

2. **The correct answer is (B).** We have been given that $t = 2$
$$(t - 4)(t + 3) = (2 - 4)(2 + 3) = -10$$

3. **The correct answer is (A).** We have $\dfrac{3}{\frac{2}{(x-1)^2}}$.

    Dividing 3 by $\dfrac{2}{(x-1)^2}$ is the same as multiplying it by $\dfrac{(x-1)^2}{2}$, which is the inverse of $\dfrac{2}{(x-1)^2}$

    Thus, we get $\dfrac{3}{\frac{2}{(x-1)^2}} = (3)\left(\dfrac{(x-1)^2}{2}\right) = \dfrac{3(x-1)^2}{2}$

4. **The correct answer is (E).** To solve this problem, we make use of the formula of "Difference of Squares." We have
$$(x + y - 2)(x + y + 2) = ((x + y) - 2)((x + y) + 2)$$
$$= (x + y)^2 - (2)^2$$
$$= (x + y)^2 - 4$$

5. **The correct answer is (D).** In this problem, we note that $9x^2$ can be written as $3(3x^2)$. Thus, our original problem becomes
$$(9x^2)^2 = (3(3x^2))^2 = (3(5))^2 = 225$$

6. **The correct answer is (C).** Looking at the figure below, we see that the coordinates of point D are (6, 1) and those of B are (1, 6). This gives us the coordinates of point C as (6, 6).

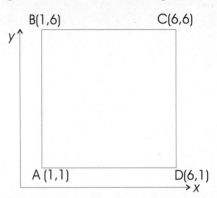

7. **The correct answer is (B).** To calculate the point of intersection of a curve with the y-axis, we note that the point of intersection has its x-coordinate $= 0$. Using this information, we plug in $x = 0$ in the equation of the curve. Thus, we have

$$3x - 2y + 6 = 0$$
$$\Rightarrow 3(0) - 2y + 6 = 0$$
$$\Rightarrow y = 3$$

8. **The correct answer is (C).** Let the number of apples in bag C be n. Then, bag B contains $\frac{n}{3}$ and bag A contains $\frac{\left(\frac{n}{3}\right)}{2}$ apples. We have been given that bag A contains 30 apples. Therefore, $\frac{\left(\frac{n}{3}\right)}{2} = 30$, which tells us that the value of $n = 180$.

9. **The correct answer is (E).** Let the number be x. Then, $(\sqrt[3]{x})^3 = 27$. Solving for x, we have $x = (27^{\frac{1}{3}})^2 = 9$.

10. **The correct answer is (B).** In this problem, we use the property of "alternate angles" for a pair of parallel lines, which states that alternate angles formed by a line intersecting a pair of parallel lines are equal. Thus, $\angle m = 60°$.

11. **The correct answer is (A).** We can solve this problem in two ways. One way is to solve the pair of equations simultaneously. Subtracting the first equation from the second equation gives us $y = 2$. Plugging back that value of y in the first equation, we get $x = 5 - y = 5 - 2 = 3$. The other way is to see that $x + 2y = 7$ can be written as $(x + y) + y = 7$.

Now, we know that $x + y = 5$. Therefore, we have $(5) + y = 7 \Rightarrow y = 2$. We can get the value of x, as in the above case.

ANSWERS AND EXPLANATIONS

12. **The correct answer is (E).** We have been given that $2x + y + 3x = 2y$. We need to calculate the value of $x + y$. Now,

    $2x + y + 3x = 2y$
    $\Rightarrow 5x + y = 2y$
    $\Rightarrow 5x = y$
    $\Rightarrow 5x + x = y + x$
    $\Rightarrow x + y = 6x$

13. **The correct answer is (B).** $f(x) = 2x - 1$. To get the value of $f(2.2)$, we plug in $x = 2.2$ for x in the above function. Thus, $f(2.2) = 2(2.2) - 1 = 3.4$.

14. **The correct answer is (C).** We are given the value of $sin\theta$. To obtain the value of $cos\theta$, we use the trigonometric identity $sin^2\theta + cos^2\theta = 1$.

    This gives us $cos^2\theta = (1 - sin^2\theta) \Rightarrow cos\theta = \sqrt{(1 - sin^2\theta)}$.

    Thus, $cos\theta = \sqrt{1 - (0.6)^2} = 0.8$.

15. **The correct answer is (A).** In this problem, we note that the left side of the equation, $2^3 \cdot 2^2$, is the same as 2^{3+2}, which is equal to 2^5. Thus, we have $2^5 = 2^k$. Equating the exponents, we have $k = 5$.

16. **The correct answer is (D).** We have been given a function $f(x)$, such that $f(x) = \dfrac{1}{x^2}$. To find $f(2x - 3)$, we plug in $(2x - 3)$ for x in the given function. Thus, $f(2x - 3) = \dfrac{1}{(2x - 3)^2}$.

17. **The correct answer is (E).** The percentage increase in income is given by the increase in income over the original income times 100. Thus, if $60 is the original income and $72 is the new income, the net increase in income is ($72 - 60) = $12.

    Therefore, the percentage increase in income is $\left(\dfrac{12}{60}\right)100 = 20\%$.

18. **The correct answer is (C).** If $|f(x)| = k$, that means either $f(x) = k$ or $-f(x) = k$. Thus, for the given problem, we have

    $|(x - 3)| > 2$
    $\Rightarrow (x - 3) > 2$ or $-(x - 3) > 2$
    $\Rightarrow x > 5$ or $-x + 3 > 2$
    $\Rightarrow x > 5$ or $-x > -1$
    $\Rightarrow x > 5$ or $x < 1$

19. **The correct answer is (D).** To calculate the cost of wrapping the box, we need to first find the total amount of wrapping paper needed to wrap the box. That will be the same as the surface area of the box. The surface area of the box is twice the sum of the area of the three rectangles that make up the box. The dimensions of the rectangles are 100×50, 100×20, and 50×20, all in cm. Thus, the surface area of the box is

$$2(100 \times 50) + 2(100 \times 20) + 2(50 \times 20) = 16{,}000 \text{ cm.sq.}$$
$$= 1.6 \text{ meter sq.}$$

The cost of wrapping would be the cost per sq. meter of wrapping times the area to be wrapped in sq. meters. Therefore, the cost of wrapping $= (1.2)(1.6) = \$1.92$.

20. **The correct answer is (C).** Let $\$x$ be the price of each peach. Then the price of 5 peaches is $\$5x$, which is given to us as $\$7.50$. Equating the two values, we have: $5x = 7.50$, which gives us $x = 1.50$. Hence, the price of 8 peaches will be $8x = 8(1.5) = \$12.00$.

21. **The correct answer is (A).** A mixture of two compounds, A and B, contains 30% A. Therefore, in a mixture that weighs 30 mg, the quantity of compound A will be 30% of 30 mg, which is equal to $0.3(30) = 9$ mg.

22. **The correct answer is (C).** We need to find the equation of line A, given that it is parallel to the line $y = 3x - 6$, and it cuts the y-axis at point $(0, 4)$. Now, if line A is parallel to a given line, it has the same slope as the line, which is 3 (comparing the equation of the line with $y = mx + b$, where m = slope of the line and b is the y-intercept). The line A cuts the y-axis at point $(0, 4)$, which means its y-intercept is 4. Thus, the equation of line A is $y = 3x + 4$.

23. **The correct answer is (D).** In this problem, we observe that triangle BCD is a right-angled triangle, with BD as the hypotenuse.

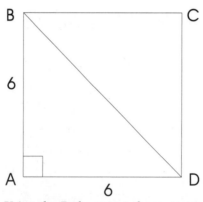

Using the Pythagorean theorem we have

$$(BD)^2 = (BC)^2 + (CD)^2 = (6)^2 + (6)^2 = 72$$
$$\Rightarrow l(BD) = \sqrt{72} = 6\sqrt{2}$$

ANSWERS AND EXPLANATIONS

24. **The correct answer is (B).** Here, we use the property of triangles, which says that the sum of all internal angles in a triangle is 180°. Also, since $AB = AC$, we know that $\angle ABC = \angle ACB = 55°$. Then, we have $\angle ABC + \angle ACB + \angle BAC = 180° \Rightarrow \angle BAC = 180 - 55 - 55 = 70°$.

25. **The correct answer is (C).** Since the regular hexagon is made up of 6 equilateral triangles, $l(AB) = l(BC) = l(CD) = l(DE) = l(EF) = l(FA) = l(OA) = 3$ cm.

    The perimeter of the hexagon $= l(AB) + l(BC) + l(CD) + l(DE) + l(EF) + l(FA) = 18$ cm.

26. **The correct answer is (A).** We draw the following figure for this problem.

    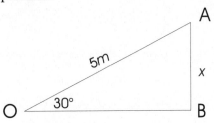

    Here, triangle OAB is a right-angled triangle, with OA as the hypotenuse. The height that the person goes to after driving 5 meters along the elevated driveway is AB, which can be obtained by using the definition of sine of an angle.

    $\sin 30 = \dfrac{AB}{OA}$

    $\Rightarrow 0.5 = \dfrac{AB}{5}$

    $\Rightarrow AB = 0.5(5) = 2.5$ m

27. **The correct answer is (D).** Since the line segment is horizontal, the y-coordinate of its end points are going to be the same as that of the midpoint. Let $(a, -2)$ and $(b, -2)$ be the end points of the line segment.

    One way to obtain the values of a and b would be to observe that since the length of the line segment is 8 and $(3, -2)$ is the midpoint, $a = (3 - \dfrac{8}{2})$ and $b = (3 + \dfrac{8}{2})$.

    Thus, we have the two end points as $(-1, -2)$ and $(7, -2)$.

    The other way to get the values of a and b would be to form a pair of equations and solve them simultaneously for a and b.

    Since $(3, -2)$ is the midpoint of $(a, -2)$ and $(b, -2)$ and the length of the line segment is 8 units, we have the following two relations between a and b

    $\dfrac{a+b}{2} = 3$, and $b - a = 8$

    Solving them simultaneously, we get $a = -1$ and $b = 7$.

28. **The correct answer is (E).** In the given circle, triangle *OAB* forms a right-angled triangle with *AB* as the hypotenuse. Since the radius of the circle is 4 units, *OA* = *OB* = 4.

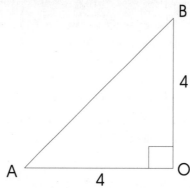

Using the Pythagorean theorem, we have

$(AB)^2 = (OA)^2 + (OB)^2 = 16 + 16 = 32$
$\Rightarrow l(AB) = \sqrt{32} = 5.656$ units

29. **The correct answer is (D).** If r is the radius of a circle and x is the angle subtended by an arc of the circle at the center, the length of that arc, *L*, is given as

$$L = \frac{x}{360}(2\pi r)$$

Thus, for a circle with radius 5, the length of an arc that subtends an angle of 45° at the center is $L = \frac{45}{360}(2\pi)(5) = 3.93$ units.

30. **The correct answer is (C).** The volume of the given solid can be obtained by calculating the volume of the cone with a base radius of 3 and a height of 20 and subtracting from it the volume of the cone base radius of 1.5 and a height of 10.

The volume of a cone base radius r and height h is given as $V = \frac{1}{3}\pi r^2 h$. Therefore, the required volume is *V* = (Volume of the bigger cone) − (Volume of the smaller cone).

$$V = \frac{1}{3}\pi(3)^2(20) - \frac{1}{3}\pi(1.5)^2(10) = 164.93$$

ANSWERS AND EXPLANATIONS

31. **The correct answer is (B).** The area of triangle ABC is $\frac{1}{2}(AC)(BC)$.

    We need to get the lengths of sides AB and BC of the triangle. It is given that triangle ABC is a right-angled triangle with AB as a diameter of the circle. That implies that $AC = BC$ and that AB is the hypotenuse of the triangle.

    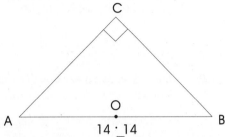

    Now, $l(AB) = 10\sqrt{2}$, since the radius = $5\sqrt{2}$. Using the Pythagorean theorem, we have

    $$(AB)^2 = (AC)^2 + (BC)^2 = 2(AC)^2$$
    $$\Rightarrow (10\sqrt{2})^2 = 2(AC)^2$$
    $$\Rightarrow l(AC) = 10 = l(BC)$$

    Therefore, the area of the triangle is $\frac{1}{2}(AC)(BC) = 50$ units.

32. **The correct answer is (E).**

    $$\sqrt{\frac{2}{3}} = \frac{\sqrt{2}}{\sqrt{3}} = \left(\frac{\sqrt{2}}{\sqrt{3}}\right)\left(\frac{\sqrt{3}}{\sqrt{3}}\right) = \frac{\sqrt{(2)(3)}}{\sqrt{(3)(3)}} = \frac{\sqrt{6}}{3}.$$

33. **The correct answer is (C).** We have a total of 10 options for the first position, 26 options for the second position, and 10 options for the third position of the combination.

    For each of the 10 options of the first position, we have 26 unique options for the second position, and for each of those options, we have 10 unique options for the third position.

    Thus, the total number of unique combinations possible are $(10)(26)(10) = 2,600$.

34. **The correct answer is (B).** An ant traverses a circular path in 12 minutes; that is, it covers a complete 360° in 12 seconds. Therefore, in 2 seconds it will cover a total of $\frac{2}{12}(360) = 60°$.

35. **The correct answer is (D).** Here, we use the definition of a tangent of an angle. We have

    $$\tan 35 = \frac{x}{12} \Rightarrow x = 12(\tan 35) = 12(0.7) = 8.40$$

36. **The correct answer is (B).** In this problem, we have been given $a(t) = 16t^2 + 100t + 75$. We need to find the value of $a(5)$, which is the acceleration after 5 seconds. Thus,

    $$a(5) = 16(5)^2 + 100(5) + 75 = 975$$

37. **The correct answer is (C).** We first multiply out the given expression and then use the trigonometric identity $sin^2\theta + cos^2\theta = 1$ to simplify our expression. Thus, we have

$$(sin\theta + cos\theta)^2 = sin^2\theta + 2sin\theta cos\theta\ cos^2\theta = 1 + 2\ sin\theta cos\theta.$$

38. **The correct answer is (D).** We have been given that triangles ABC and CED are similar. By the property of similar triangles, we have

$$\frac{AB}{CE} = \frac{BC}{ED} = \frac{AC}{CD} \Rightarrow \frac{x}{y} = \frac{3}{4}.$$

39. **The correct answer is (D).** Since $f(x) = (1 + 2x)^2$ is a square function, the least value it can take is a zero, it does take this value at $x = -\frac{1}{2}$. Also, for $x = 2$, the function takes the value $f(2) = (1 + 2 \cdot 2)^2 = 25$.

40. **The correct answer is (A).** The shortest distance between the two towns can be found using the distance formula, which gives the shortest distance between two points $(a1, b1)$ and $(a2, b2)$ as

$$D = \sqrt{(a2 - a1)^2 + (b2 - b1)^2}.$$

Thus, the distance between the two towns having coordinates (3, 5) and (8, 11) using the distance formula is

$$D = \sqrt{(8 - 3)^2 + (11 - 5)^2} = \sqrt{61}.$$

41. **The correct answer is (B).** This problem can be solved using the formula of exponential growth, which is $A = P(1 + r)^t$, where A = the amount present finally, P = the initial amount, r = the rate of growth, and t = the total time.

Thus, for our problem, we have

$$A = 5,000(1 + 0.04)^{10} = 7,401 \approx 7,400.$$

42. **The correct answer is (D).** The sum of all internal angles of a quadrilateral is 360°. This gives us $p + q + r + s = 360$.

Now, if $(r + s) = 360 - (p + q)$ or $p = 70$ and $90 < q < 150$.

If $q = 90$, then $(r + s) = 360 - (70 + 90) = 200$, and

If $q = 150$, then $(r + s) = 360 - (70 + 150) = 140$.

Thus, we see that the value of $(r + s)$ lies between 140 and 200.

43. **The correct answer is (E).** The first bucket gets 1 marble. The second bucket gets twice of the first one, which is 2^{2-1} marbles. The third bucket gets 2(2) marbles, which is 2^{3-1} marbles. The fourth bucket gets 8 marbles, which is the same as 2^{4-1}. Thus, we observe that the nth bucket will get 2^{n-1} marbles.

ANSWERS AND EXPLANATIONS

44. **The correct answer is (A).** In this problem, by symmetry, we have $AX = BX$, which means that triangle AXB is an isosceles triangle. Thus, if the perimeter of the triangle is 36 and $AB = 10$ (since AB is the diameter of the circle of radius 5), then we have $AX + BX + AB = 36$, which gives us $AX = BX = 13$.

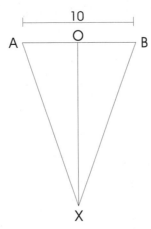

Now, to find the height of the cylinder, we observe that triangle OXB is a right-angled triangle. Using the Pythagorean theorem, we have $OX = 12$.

Thus, the height of the cylinder is 12.

45. **The correct answer is (C).** To find $f(g(t))$, we simply plug in $g(t) = t + 1$ for x in the function $f(x) = x^2 + 2x - 1$. Thus, $f(g(t)) = f(t + 1) = (t + 1)^2 + 2(t + 1) - 1 = t^2 + 4t + 2$.

46. **The correct answer is (D).** There are a total of $(3 + 4 + 5) = 12$ balls. Of these, there are a total of $(3 + 4) = 7$ white or red balls. Thus, the probability of drawing a red or white ball is $\frac{7}{12}$.

47. **The correct answer is (B).** The equations of asymptotes to a hyperbola $\frac{x^2}{a^2} - \frac{y^2}{b^2} = 1$ are $y = \frac{b}{a}x$ and $y = -\frac{b}{a}x$.

Thus, the equations of asymptotes to the hyperbola $\frac{x^2}{36} - \frac{y^2}{25} = 1$ are $y = \frac{5}{6}x$ and $y = -\frac{5}{6}x$.

48. **The correct answer is (B).** The mean of a set of numbers is defined as the average value of the set, whereas the median of a set of numbers is defined as the value of the number in the middle of the set. If any one number in the set is changed, the value of the mean changes, but the median need not.

In this case, since the number with the greatest value is increased, the mean changes, but the median remains the same.

49. **The correct answer is (B).** Since $sin^2\theta + cos^2\theta = 1$, the problem reduces to $(1 - 4)^2 = 9$. Here, we notice that the value of $sin\theta = 0.63$ has been given to mislead us.

50. **The correct answer is (D).** Using the property of exponents, we have $(a^2)^3 + (b^2)(b^3) = a^{2(3)} + b^{2+3} = a^6 + b^5$.

Practice Test 2

MATHEMATICS LEVEL IC TEST

PRACTICE TEST 2

MATHEMATICS LEVEL IC TEST

While you have taken many standardized tests and know to blacken completely the ovals on the answer sheets and to erase completely any errors, the instructions for the SAT II Mathematics IC exam differs in three important ways from the directions for other standardized tests you have taken. You need to indicate on the answer key which test you are taking.

The instructions on the answer sheet will tell you to fill out the top portion of the answer sheet exactly as shown.

1. Print MATHEMATICS LEVEL IC on the line to the right under the words *Subject Test (print)*.

2. In the shaded box labeled *Test Code* fill in four ovals:

   —Fill in oval 3 in the row labeled V.
   —Fill in oval 2 in the row labeled W.
   —Fill in oval 5 in the row labeled X.
   —Fill in oval A in the row labeled Y.
   —Leave the ovals in row Q blank.

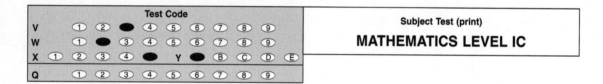

When everyone has completed filling in this portion of the answer sheet, the supervisor will tell you to turn the page and begin the Mathematics Level IC examination. The answer sheet has 100 numbered ovals on the sheet, but there are only 50 multiple-choice questions in the test, so be sure to use only ovals 1 to 50 to record your answers.

MATHEMATICS LEVEL IC TEST

REFERENCE INFORMATION

The following information is for your reference in answering some of the questions in this test.

Volume of a right circular cone with radius r and height h:
$$V = \frac{1}{3}\pi r^2 h$$

Lateral area of a right circular cone with circumference of the base c and slant height l:
$$S = \frac{1}{2}cl$$

Surface area of a sphere with radius r:
$$S = 4\pi r^2$$

Volume of a pyramid with base area B and height h:
$$V = \frac{1}{3}Bh$$

For each of the following problems, identify the BEST answer of the choices given. If the exact numerical value is not one of the choices, select the answer that is closest to this value. Then fill in the corresponding oval on the answer sheet.

NOTES

1. You will need a calculator to answer some (but not all) of the questions. You must decide whether or not to use a calculator for each question. You must use at least a scientific calculator; you are permitted to use graphing or programmable calculators.

2. Degree measure is the only angle measure used on this test. Be sure your calculator is set in degree mode.

3. The figures that accompany questions on the test are designed to give you information that is useful in solving problems. They are drawn as accurately as possible EXCEPT when stated that a figure is not drawn to scale. Unless otherwise indicated, all figures lie in planes.

4. The domain of any function f is assumed to be the set of all real numbers x for which $f(x)$ is a real number, unless otherwise specified.

PRACTICE TEST 2

MATHEMATICS LEVEL IC TEST

USE THIS SPACE FOR SCRATCH WORK

1. Solve the system of equations $x + y = 2$ and $2x - 3y = 1$.

   (A) $\left(\frac{7}{5}, \frac{3}{5}\right)$
   (B) $(3, 7)$
   (C) $(5, -3)$
   (D) $(7, 3)$
   (E) $\left(\frac{3}{5}, \frac{7}{5}\right)$

2. If $f(x) = 2x$, which of the following is true?

   (A) $f(-a) = -f(a)$
   (B) $f(-a) = f(a)$
   (C) $f(ab) = f(a)f(b)$
   (D) $f\left(\frac{a}{b}\right) = \frac{f(a)}{f(b)}$
   (E) both choices (A) and (C)

3. A triangle has a base of 8 units and a height of 6 units. A second triangle with a base of 4 units and a height of 3 units overlaps the first triangle as shown to the right. What is the difference in area between the two nonoverlapping regions of the two triangles?

   (A) 24 square units
   (B) 6 square units
   (C) 18 square units
   (D) 12 square units
   (E) Cannot be determined with the given information

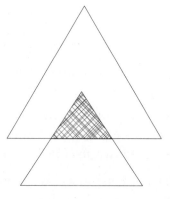

Peterson's SAT II Success:
Mathematics IC and IIC

MATHEMATICS LEVEL IC TEST—Continued

USE THIS SPACE FOR SCRATCH WORK

4. If the vertex of a parabola $f(x)$ is at (1, 1), where is the vertex of $f(x - 1) - 2$?

   (A) (0, −1)
   (B) (0, 3)
   (C) (2, 3)
   (D) (2, −1)
   (E) (1, 1)

5. The expression $\dfrac{a^{-1} + b^{-1}}{a^{-1}}$ is equivalent to

   (A) b^{-1}
   (B) $\dfrac{a + b}{a}$
   (C) $\dfrac{a + b}{b}$
   (D) $\dfrac{b}{(a + b)}$
   (E) $\dfrac{a}{(a + b)}$

6. If the discriminant of a quadratic equation equals −4, the equation has how many real roots?

   (A) 0
   (B) 1
   (C) 2
   (D) 3
   (E) −2

GO ON TO THE NEXT PAGE

MATHEMATICS LEVEL IC TEST—Continued

7. Given the figure to the right, find the value of x.

   (A) -2
   (B) -2 and 2
   (C) 2
   (D) 8
   (E) none of the above

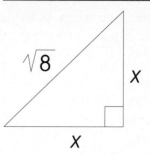

8. If $(a + 3i)(2 + i) = 7 + 11i$, what is the value of a?

   (A) 2
   (B) 5
   (C) 3
   (D) -5
   (E) -2

9. How many ways can a 10-question true-and-false exam be answered if no questions are left blank?

   (A) 512
   (B) 100
   (C) $1,024$
   (D) 10
   (E) $3,628,800$

10. What is the y-intercept of $3y - 2x + 6 = 0$?

    (A) $(0, 3)$
    (B) $(0, 6)$
    (C) $(0, 2)$
    (D) $(2, 0)$
    (E) $(0, -2)$

11. The area of a circle whose radius extends from (−2, 5) to (5, 1) is

    (A) 11π
    (B) 25π
    (C) 45π
    (D) 65π
    (E) 9π

12. If $f(x) = 2x + b$ and $f(-2) = 3$, b equals

    (A) 7
    (B) −1
    (C) 1
    (D) −8
    (E) −7

13. If $x^3 - 2x^2 + ax + 9$ is exactly divisble by $x + 3$, a equals

    (A) −10
    (B) −12
    (C) 7
    (D) 16
    (E) 10

14. The slope of the line $3x - 2y = 6$ is

    (A) 3
    (B) −3
    (C) $\dfrac{3}{2}$
    (D) $\dfrac{-3}{2}$
    (E) $\dfrac{2}{3}$

MATHEMATICS LEVEL IC TEST—*Continued*

USE THIS SPACE FOR SCRATCH WORK

15. A swimming pool is 40 feet long and 4 feet deep at one end. If it is 8 feet deep at the other end, find the total distance in feet along the bottom.

    (A) 40
    (B) 40.20
    (C) 44
    (D) 40.79
    (E) 40.15

16. Which of the following functions is one to one?

    (A) $f(x) = x^2 + 1$
    (B) $f(x) = 3$
    (C) $f(x) = \sqrt{x}$
    (D) $f(x) = x^4$
    (E) $\sin x$

17. The solution set to $x^2 - 6 < x$ is

    (A) $(-2, 3)$
    (B) all real numbers except the interval $(-2, 3)$
    (C) all real numbers except the interval $(-3, 2)$
    (D) $(-3, 2)$
    (E) $(-2, 3)$

18. If Joe can paint a house in 6 hours and Jim can paint a house that is exactly the same size in 5 hours, how many hours does it take if both work together?

    (A) 5.5
    (B) 5
    (C) 3.27
    (D) 2.52
    (E) 2.73

19. The four points A(3, 3), B(7, 3), C(5, 5), and D(9, 5) are the vertices of a parallelogram. The points A′, B′, C′, and D′ are found by multiplying each of the x coordinates of A, B, C, and D, respectively, by -1. Find the difference between the areas of parallelogram ABCD and parallelogram A′B′C′D′.

   (A) 15
   (B) 6
   (C) 4
   (D) 10
   (E) 0

20. Given rectangle ABCD in the figure to the right with $AB = 12$, $BC = 9$, $EB = BF = 3$, find the area of triangle DEF.

   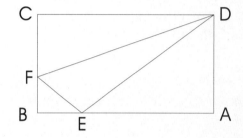

   (A) $\dfrac{45}{2}$
   (B) 27
   (C) 35
   (D) $\dfrac{83}{2}$
   (E) 51

21. The expression $\dfrac{\sqrt{2}+1}{\sqrt{2}-1}$ is equivalent to

   (A) 3
   (B) -1
   (C) $5 + \sqrt{2}$
   (D) $3 + 2\sqrt{2}$
   (E) 1

22. Which statement illustrates the commutative property of multiplication?

   (A) $a(b + c) = (b + c)a$
   (B) $(ab)c = a(bc)$
   (C) $a(0) = 0$
   (D) $a(b + c) = ab + ac$
   (E) $a(1) = a$

23. What is the range of $\cos x$?

   (A) all real numbers
   (B) $(-\infty, \infty)$
   (C) $(-1, 1)$
   (D) $(-1, 1)$
   (E) $(0, \infty)$

24. If $f(x) = x^3 + 1$, then $f^{-1}(9)$ is

   (A) 730
   (B) 2
   (C) $\sqrt[3]{3}$
   (D) $\sqrt[3]{10}$
   (E) 1

25. Two circles, one inside the other, are centered at the same point and are 5 units apart. What is the difference between the circumferences of the circles?

   (A) $\dfrac{5\pi}{2}$
   (B) 5π
   (C) 10π
   (D) 15π
   (E) Cannot be determined from the given information

26. If $x = 2^a$, then 2^{a+3} equals

   (A) $x + 3$
   (B) x^3
   (C) $6x$
   (D) $8x$
   (E) $3x$

27. If $\sin x = \dfrac{3}{5}$ and $\cos x < 0$, then find $\tan x$.

   (A) $-\dfrac{4}{3}$
   (B) $\dfrac{3}{4}$
   (C) $-\dfrac{3}{4}$
   (D) $\dfrac{4}{3}$
   (E) $\dfrac{4}{5}$

28. Convert the binary number 101,101 to a base 10 number.

   (A) 45
   (B) 18
   (C) 46
   (D) 22
   (E) 90

MATHEMATICS LEVEL IC TEST—Continued

29. Given 7 dots, no 3 of which are on a straight line, how many line segments need to be drawn to connect every pair of dots with a straight line?

   (A) 21
   (B) 42
   (C) 24
   (D) 32
   (E) 28

30. Find x for the triangle in the figure to the right.

   (A) 25
   (B) 35
   (C) 20
   (D) 30
   (E) 15

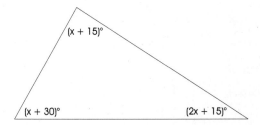

31. How many triangles are in the figure to the right?

   (A) 7
   (B) 9
   (C) 17
   (D) 19
   (E) 15

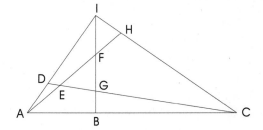

32. The equation of the line that passes through the point $(-3, 1)$ and is perpendicular to $2y - x = 8$ is

   (A) $4y - 2x = -1$
   (B) $2y = x + 5$
   (C) $y = 2x + 7$
   (D) $y + 2x + 5 = 0$
   (E) $y - 7 = -2x$

33. The reciprocal of i is

(A) i
(B) 1
(C) -1
(D) $\sqrt{-1}$
(E) $-i$

34. What is the cosine of 60 degrees?

(A) $\dfrac{\sqrt{3}}{2}$
(B) $\dfrac{3}{2}$
(C) $\dfrac{1}{2}$
(D) $-\dfrac{1}{2}$
(E) $-\dfrac{\sqrt{3}}{2}$

35. If $f(x) = 2x + 1$ and $g(x) = x - 1$, then what is the domain of $\left(\dfrac{f}{g}\right)(x)$?

(A) all real numbers
(B) all real numbers except $\left(-\dfrac{1}{2}, \dfrac{1}{2}\right)$
(C) all real numbers except $(-1, 1)$
(D) all real numbers except 1
(E) all real numbers except $(-1, 1)$

MATHEMATICS LEVEL IC TEST—Continued

36. If the mode of a list of numbers is 10 and we add distinct numbers 7, 13, 18 to the list, then what number is guaranteed to be a mode of the new list?

(A) 11
(B) 18
(C) 7
(D) 13
(E) 10

37. The figure to the right shows three of the faces of a cube. If the six faces of the cube are numbered consecutively, what could the sum of the numbers on all six faces be?

I. 240
II. 237
III. 243

(A) I only
(B) II only
(C) III only
(D) I or II
(E) II or III

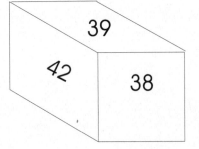

38. What is x in degrees where x is given in the figure to the right?

(A) 60
(B) 45
(C) 30
(D) 20
(E) 25

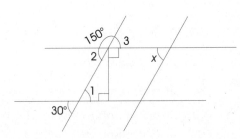

39. If the width of a rectangle is one fourth the length and the perimeter is 70 cm, what is the area?

    (A) 140 square cm
    (B) 196 square cm
    (C) 392 square cm
    (D) 784 square cm
    (E) 70 square cm

40. If two positive consecutive integers have a product of 342, then what are the integers?

    (A) 18 and 19
    (B) 19 and 20
    (C) 17 and 18
    (D) 2 and 171
    (E) 9 and 19

41. Given that the foci of the ellipse are the points (3, 0) and (−3, 0), then find a where $16x^2 + (ay)^2 = (4a)^2$.

    (A) 5
    (B) −5
    (C) 4
    (D) −4
    (E) 3

MATHEMATICS LEVEL IC TEST—Continued

USE THIS SPACE FOR SCRATCH WORK

42. Wanting to buy a pair of shoes, Bob has a choice. Two competing stores next to one another carry the same brand of shoes with the same list price of $100 but at two different discount schemes. Shoe Store A offers a 20 percent discount year-round for shoes more than 60 dollars, but on this particular day it offers an additional discount of 10 percent off their already discounted price. Shoe Store B simply offers a discount of 30 percent on that same day in order to stay competitive. Bt what percentage of list price do the two options differ?

   (A) 2
   (B) 0
   (C) 1
   (D) 10
   (E) 3

43. Given the graphs of $f(x)$ and $g(x)$ to the right, what is $(f \cdot g)(2)$?

   (A) 3
   (B) 5
   (C) 2
   (D) 6
   (E) −2

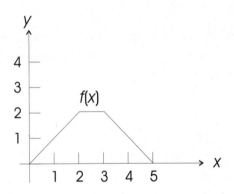

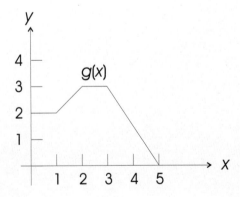

44. A total of 8 students took a test, and their average score was 77. If the average score of 4 of the students was 73, what was the average score for the remaining 4 students?

(A) 81
(B) 80
(C) 79
(D) 78
(E) 77

45. In the right triangle (to the right), AB is twice BC. What is the length of BC?

(A) 125^2
(B) $5\sqrt{5}$
(C) $\sqrt{5}$
(D) 625
(E) $\sqrt{25}$

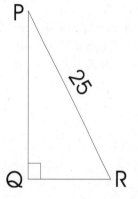

46. Find the difference between the areas of the two right triangles ABC and BCD in the figure to the right where $BC = 12$, $BD = 20$, and $AD = 6$.

(A) 36
(B) 60
(C) 96
(D) 132
(E) 72

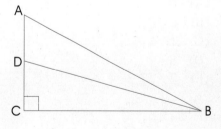

MATHEMATICS LEVEL IC TEST—Continued

47. In the figure below, *ABC* is an equilateral triangle with *KLPQ* and *MLPN* as two trapezoids. What is the ratio of the area covered by both trapezoids to the area of the triangle *ABC*, if $AK = KL = LB = \frac{AB}{3}$?

    (A) 1:4
    (B) 2:5
    (C) 2:3
    (D) 2:9
    (E) 1:3

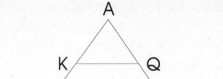

48. The surveyor needs to calculate the length of a pond. He drew the diagram to the right and made measurements as indicated. What is the length of the pond?

    (A) 100
    (B) 96
    (C) 90
    (D) 120
    (E) 140

49. Given the figure to the right, find *y*.

    (A) 0
    (B) 2
    (C) 1
    (D) 8
    (E) 7

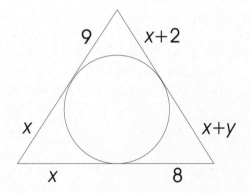

MATHEMATICS LEVEL IC TEST—Continued

USE THIS SPACE FOR SCRATCH WORK

50. If $\theta = \arccos\left(\dfrac{u}{v}\right)$, then $\tan \theta$ is

   (A) $\dfrac{\sqrt{v^2 - u^2}}{u}$

   (B) $\dfrac{u}{v}$

   (C) $\dfrac{\sqrt{v^2 - u^2}}{u}$

   (D) $\dfrac{v}{u}$

   (E) $\sqrt{v^2 - u^2}$

STOP If you finish before time is called, you may check your work on this section only. Do not turn to any other section in the test.

Quick Score Answers

1. A	11. D	21. D	31. C	41. A
2. A	12. A	22. A	32. D	42. A
3. C	13. B	23. D	33. E	43. C
4. D	14. C	24. B	34. C	44. A
5. C	15. B	25. C	35. D	45. B
6. A	16. C	26. D	36. E	46. A
7. C	17. A	27. C	37. E	47. C
8. B	18. E	28. A	38. C	48. D
9. C	19. E	29. A	39. B	49. C
10. E	20. B	30. D	40. A	50. C

ANSWERS AND EXPLANATIONS

1. **The correct answer is (A).** Solving the equation $x + y = 2$ for x and substituting it into $2x - 3y = 1$, we get $y = \frac{3}{5}$ and so $x = \frac{7}{5}$.

2. **The correct answer is (A).** We see that $f(-a) = 2(-a) = -2a = -f(a)$, and so $f(-a)$ is equal to $-f(a)$ but $f(-a)$ is not equal to $f(a)$. Also, $f(ab) = 2ab$, which is not equal to $f(a)f(b) = 2a(2b) = 4ab$. Similarly $f\left(\frac{a}{b}\right) = 2\left(\frac{a}{b}\right)$, which is not equal to $\frac{f(a)}{f(b)} = \frac{2a}{2b} = \frac{a}{b}$.

3. **The correct answer is (C).** Let A_1 be the area of the 8-by-6 triangle, A_2 be the area of the 4-by-3 triangle, and A be the area of the overlap. The difference between the two nonoverlapping areas is

$$A_d = (A_1 - A) - (A_2 - A)$$
$$= A_1 - A - A_2 + A$$
$$= A_1 - A_2$$
$$= \frac{(8(6))}{2} - \frac{(4(3))}{2}$$
$$= 24 - 6$$
$$= 18.$$

Notice the area overlap does not matter.

4. **The correct answer is (D).** The graph of $f(x - 1) - 2$ is just the graph of $f(x)$ shifted to the right 1 unit and down 2 units, so the vertex of $f(x - 1) - 2$ is just the vertex of $f(x)$ shifted to the right 1 unit and down 2 units. Thus, the vertex of $f(x - 1) - 2$ is $(2, -1)$.

5. **The correct answer is (C).** Looking at the numerator, we see that it simplifies to $a^{-1} + b^{-1} = \frac{1}{a} + \frac{1}{b} = \frac{a+b}{ab}$. So, the whole fraction simplifies to

$$\frac{a^{-1} + b^{-1}}{a^{-1}} = \frac{\frac{a+b}{ab}}{\frac{1}{a}} = \frac{a(a+b)}{ab} = \frac{a+b}{b}.$$

6. **The correct answer is (A).** If the discriminant equals -4, then there are two complex roots. There are is no real roots.

7. **The correct answer is (C).** Using the Pythagorean theorem we have $x^2 + x^2 = 8$. So, $2x^2 = 8$, which says that $x^2 = 4$. Thus, $x = \pm 2$. Since $x > 0$, $x = 2$.

8. **The correct answer is (B).** $(a + 3i)(2 + i) = 2a + 6i + ai + 3i^2 = (2a - 3) + (6 + a)i$. Since this is equal to $7 + 11i$, $2a - 3 = 7$ and $6 + a = 11$. Hence, $a = 5$.

9. **The correct answer is (C).** For each question there are 2 choices and 10 questions. So, there are $2(2)(2)(2)(2)(2)(2)(2)(2)(2) = 2^{10} = 1,024$ total choices.

10. **The correct answer is (E).** Writing the equation in standard form, $y = \left(\frac{2}{3}\right)x - 2$ we see the y-intercept is $(0, -2)$.

11. **The correct answer is (D).** The distance from $(-2, 5)$ to $(5, 1)$ is $r = \sqrt{(-2-5)^2 + (5-1)^2} = \sqrt{65}$. So, $A = \pi(\sqrt{65})^2 = 65\pi$.

12. **The correct answer is (A).** $f(-2) = 3$ implies that $3 = 2(-2) + b$. So, $b = 7$.

13. **The correct answer is (B).** If $x + 3$ is a factor, then -3 is a root. So $(-3)^3 - 2(-3)^2 + a(-3) + 9 = 0$. Or, $-36 - 3a = 0$, yielding $a = -12$.

14. **The correct answer is (C).** Writing the equation of the line in slope-intercept form, we have $y = \left(\frac{3}{2}\right)x - 3$. So the slope is $\frac{3}{2}$.

15. **The correct answer is (B).**

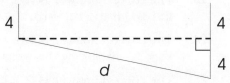

The diagram shows the picture that needs to be drawn. We need to find d. We see that d is equal to

$d = \sqrt{40^2 + 4^2} = 40.20.$

16. **The correct answer is (C).** All pass the vertical line test, but choices (A), (B), and (D) do not pass the horizontal line test. Only choice (C) does. So choice (C) is the only one-to-one function.

PRACTICE TEST 2

17. **The correct answer is (A).** $x^2 - 6 < x$ is equivalent to $x^2 - x - 6 < 0$ and $x^2 - x - 6 = (x - 3)(x + 2) < 0$, so thinking of the graph of the quadratic $x^2 - x - 6$, which opens up with x-intercepts of $x = 3$ and $x = -2$, we see that $x^2 - x - 6$ is less than 0 when x is in the interval $(-2, 3)$.

18. **The correct answer is (E).** Joe can complete 1 house in 6 hours, which can be rewritten as $\frac{1}{6}$ house per hour, and for Jim, we have $\frac{1}{5}$ house per hour. The equation for both of them together can be written as $\frac{1}{6} + \frac{1}{5} = \frac{1}{x}$.

    So, $x = \frac{6(5)}{6 + 5} = 2.73$ hours.

19. **The correct answer is (E).** By multiplying the x-coordinates by -1 we see that the parallelogram $A'B'C'D'$ is just the mirror image of parallelogram $ABCD$ about the y-axis. So there is no difference in the areas.

20. **The correct answer is (B).** The areas of the triangles BFE, FCD, and DAE are $\frac{9}{2}$, 36, and $\frac{81}{2}$, respectively. The area of DEF is just the area of rectangle $ABCD$ minus the sum of the three areas of the above triangles. So, the area of DEF is just $12(9) - \left(\frac{9}{2} + 36 + \frac{81}{2}\right) = 108 - 81 = 27$.

21. **The correct answer is (D).** Multiply both numerator and denominator by the conjugate of $\sqrt{2} - 1$, which is $-\sqrt{2} - 1$. So,

    $$\frac{\sqrt{2} + 1}{\sqrt{2} - 1} \cdot \frac{-\sqrt{2} - 1}{-\sqrt{2} - 1} = \frac{-3 - 2\sqrt{2}}{-1} = 3 + 2\sqrt{2}.$$

22. **The correct answer is (A).** The commutative property of multiplication says that you can switch the order of multiplication of two numbers. Given a as one number and $b + c$ as the other number, we have that $a(b + c) = (b + c)a$, which is choice (A). Now choice (B) is associativity, choice (C) is the property of multiplication by 0, choice (D) is the distributive property, and choice (E) is the identity for multiplication.

23. **The correct answer is (D).** We have the identity from trigonometry that says that $-1 \leq \cos x \leq 1$, so we see that $[-1, 1]$ is the range.

24. **The correct answer is (B).** $f^{-1}(9)$ says that $f(x) = 9$, which says that $x^3 + 1 = 9$. Solving this we get $x = 2$.

25. **The correct answer is (C).** If r is the radius of the inside circle, $r + 5$ is the radius of the outer circle. Their circumfrences are $2\pi r$ and $2\pi(r + 5)$, and the difference is $2\pi r + 10\pi - 2\pi r$ or 10π.

26. **The correct answer is (D).** $2^{a+3} = 2^a \times 2^3 = x \times 8 = 8x$.

ANSWERS AND EXPLANATIONS

27. **The correct answer is (C).** Since $\sin x > 0$ and $\cos x < 0$, we know that x is an angle in the second quadrant. Label a triangle with angle x and $\sin x = \frac{3}{5}$. Use the Pythagorean theorem to calculate the other side, which happens to be 4 (remember to label on the triangle -4), and read off that $\tan x = -\frac{3}{4}$.

28. **The correct answer is (A).**
$$101{,}101 = 1 \cdot 2^5 + 0 \cdot 2^4 + 1 \cdot 2^3 + 1 \cdot 2^2 + 0 \cdot 2^1 + 1 \cdot 2^0$$
$$= 32 + 8 + 4 + 1$$
$$= 45.$$

29. **The correct answer is (A).** The correct answer is 7. Choose $2 = {}_7C_2 = \frac{7 \cdot 6}{2 \cdot 1} = 21$

30. **The correct answer is (D).** The sum of the angles of a triangle is 180 degrees. So, $4x + 60 = 180$. Solving this yields $x = 30$.

31. **The correct answer is (C).** The line segments cut the figure into 7 undivided pieces. Four triangels use a single piece, 6 triangels use two pieces, 3 triangles use three pieces, 3 triangles use four pieces, and 1 triangle uses all seven pieces. $4 + 6 + 3 + 3 + 1 = 17$.

32. **The correct answer is (D).** The slope of the line that is perpendicular to $2y - x = 8$ is -2. Using the point-slope formula, we have $y - 1 = -2(x + 3)$, which yields $y = -2x - 5$ or the equation in choice (D).

33. **The correct answer is (E).** $\frac{1}{i} = \frac{1}{i} \times \frac{i}{i} = -i$.

34. **The correct answer is (C).** Thinking about the special angle, we get that cosine of 60 degrees is $\frac{1}{2}$.

35. **The correct answer is (D).** $\left(\frac{f}{g}\right)(x) = \frac{2x+1}{x-1}$. So the domain of this function is all real numbers, except where the denominator is equal to 0, which is at 1.

36. **The correct answer is (E).** If we add distinct numbers (different numbers from all the others), this does not change the mode since the mode is the number that is present the most times in the given list of numbers.

37. **The correct answer is (E).** We know by looking at the numbers on the cube that five numbers are definitely on the cube. That is, the numbers 38, 39, 40, 41, and 42 are present on the cube. We do not know if the sixth number is 37 or 43. So the two possible sums are choice (E).

38. **The correct answer is (C).** Starting at the 30-degree angle, we know by vertical angles that $\angle 1$ is 30°. This says that by alternate interior angles that angle 2 is 30, which by vertical angles $\angle 3$ is 30°. Finally, by alternate interior angles, $x = 30$.

39. The correct answer is (B). We set up the two equations, $2w + 2l = 70$ and $w = \frac{1}{4}l$, where l is the length and w is the width. Now, substituting $w = \frac{1}{4}l$ into the other equation we get $l = 28$ and so $w = 7$. Hence, the area is $28(7) = 196$ square cm.

40. The correct answer is (A). Set up the equation $n(n + 1) = 342$. So, $n(n + 1) - 342 = n^2 + n - 342 = (n + 19)(n - 18) = 0$ when $n = 18$ (n cannot be equal to -19), so the two integers are 18 and 19.

41. The correct answer is (A). First, rewrite the equation into $\frac{x^2}{a^2} + \frac{y^2}{16} = 1$. We know from the given information about the foci that $c = 3$ and from the above equation that $b = 4$. Also, $a^2 = b^2 + c^2$, so $a = 5$.

42. The correct answer is (A). Store A's first discount of 20 percent gives a reduced price of 80 dollars and then 10 percent on top of that gives a discount price of 72 dollars. Store B offers 30 percent on the 100 dollars, which gives a discount price of 70 dollars. So, the difference is 2 dollars or 2 percent.

43. The correct answer is (C). Using the graphs $(f \circ g)(2) = f(g(2)) = f(3) = 2$.

44. The correct answer is (A). If $a1, a2, a3, \ldots, a8$ are the eight scores on the test, we know that the average for all eight is given by the equation $\frac{(a1 + a2 + a3 + a4 + a5 + a6 + a7 + a8)}{8} = 77$ and say the first four have an average of $\frac{(a1 + a2 + a3 + a4)}{4} = 73$.

So, these two equations can be written as

$a1 + a2 + a3 + a4 + a5 + a6 + a7 + a8 = 8(77) = 616$, and $a1 + a2 + a3 + a4 = 4(73) = 292$.

So, $a5 + a6 + a7 + a8 = 616 - 292 = 324$, and the average is $\frac{324}{4} = 81$.

45. The correct answer is (B). Let length of $QR = x$. Then the length of $PQ = 2x$. Using the Pythagorean theorem we have

$$x^2 + (2x)^2 = 25^2$$
$$x^2 + 4x = 625$$
$$5x^2 = 625$$
$$x^2 = 125$$
$$x = \sqrt{125} = \sqrt{25 \times 5} = 5\sqrt{5}$$

46. The correct answer is (A). There are two things to notice. The difference is the area of ABD and the altitude of ABD is just BC. So the area of the difference is $\frac{6(12)}{2} = 36$.

ANSWERS AND EXPLANATIONS

47. The correct answer is (C). Just count the number of equilateral triangles in each trapezoid, which is 3. So, there are 6 total equilateral triangles in the trapezoids and 9 total. So, the ratio is 2:3.

48. The correct answer is (D). Set up congruent triangles. Then set up the ratios of sides to get 80 is to 20 as x is to 30. We see that since these triangles have to be congruent that x is 120.

49. The correct answer is (C). Since the line segments are tangent to the circle, $x + 2 = 9$, which says that $x = 7$, and $x + y = 7 + y = 8$, which says $y = 1$.

50. The correct answer is (C). If $\arccos(\frac{u}{v}) = \theta$ then $\cos\theta = \frac{u}{v}$. So, label a triangle with angle θ and $\cos\theta = \frac{u}{v}$. Find the opposite side of angle θ, which is $\sqrt{v^2 - u^2}$. Now, just read off the triangle $\tan\theta$ to be choice (C).

Practice Test 1

MATHEMATICS LEVEL IIC TEST

PRACTICE TEST 1

MATHEMATICS LEVEL IIC TEST

While you have taken many standardized tests and know to blacken completely the ovals on the answer sheets and to erase completely any errors, the instructions for the SAT II Mathematics IIC exam differs in three important ways from the directions for other standardized tests you have taken. You need to indicate on the answer key which test you are taking.

The instructions on the answer sheet will tell you to fill out the top portion of the answer sheet exactly as shown.

1. Print MATHEMATICS LEVEL IIC on the line to the right under the words *Subject Test (print)*.

2. In the shaded box labeled *Test Code* fill in four ovals:

   —Fill in oval 5 in the row labeled V.
   —Fill in oval 3 in the row labeled W.
   —Fill in oval 5 in the row labeled X.
   —Fill in oval E in the row labeled Y.
   —Leave the ovals in row Q blank.

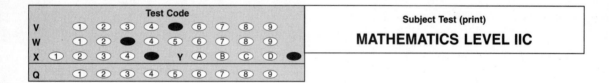

When everyone has completed filling in this portion of the answer sheet, the supervisor will tell you to turn the page and begin the Mathematics Level IIC examination. The answer sheet has 100 numbered ovals on the sheet, but there are only 50 multiple-choice questions in the test, so be sure to use only ovals 1 to 50 to record your answers.

MATHEMATICS LEVEL IIC TEST

REFERENCE INFORMATION

The following information is for your reference in answering some of the questions in this test.

Volume of a right circular cone with radius r and height h:
$$V = \frac{1}{3}\pi r^2 h$$

Lateral area of a right circular cone with circumference of the base c and slant height l:
$$S = \frac{1}{2}cl$$

Surface area of a sphere with radius r:
$$S = 4\pi r^2$$

Volume of a pyramid with base area B and height h:
$$V = \frac{1}{3}Bh$$

For each of the following problems, identify the BEST answer of the choices given. If the exact numerical value is not one of the choices, select the answer that is closest to this value. Then fill in the corresponding oval on the answer sheet.

NOTES

1. You will need a calculator to answer some (but not all) of the questions. You must decide whether or not to use a calculator for each question. You must use at least a scientific calculator; you are permitted to use graphing or programmable calculators.

2. Degree measure is the only angle measure used on this test. Be sure your calculator is set in degree mode.

3. The figures that accompany questions on the test are designed to give you information that is useful in solving problems. They are drawn as accurately as possible EXCEPT when stated that a figure is not drawn to scale. Unless otherwise indicated, all figures lie in planes.

4. The domain of any function f is assumed to be the set of all real numbers x for which $f(x)$ is a real number, unless otherwise specified.

Peterson's: www.petersons.com

MATHEMATICS LEVEL IIC TEST

1. If $\left(2 - \dfrac{x}{3}\right) = \left(1 - \dfrac{x}{2}\right)$, then $\left(1 + \dfrac{x}{6}\right) =$

   (A) -2
   (B) 0
   (C) $\dfrac{2}{3}$
   (D) $\dfrac{1}{3}$
   (E) $-\dfrac{1}{6}$

2. $(2 - 5i) + (4 + 3i) =$

   (A) $4i$
   (B) 4
   (C) $6 - 2i$
   (D) $6i - 2$
   (E) 0

3. $\dfrac{1}{x} + \dfrac{1}{y} + \dfrac{1}{z} =$

   (A) $\dfrac{xy + yz + xz}{xyz}$
   (B) $\dfrac{xy + 1}{xyz}$
   (C) $\dfrac{xyz}{x + y + z}$
   (D) $\dfrac{3}{x + y + z}$
   (E) $\dfrac{3}{xyz}$

MATHEMATICS LEVEL IIC TEST—Continued

USE THIS SPACE FOR SCRATCH WORK

4. $(a^2)^3 + \dfrac{b^3 \cdot b^2}{b^{-4}} =$

   (A) $a^5 + b$
   (B) $a^6 + b^{10}$
   (C) $a^5 + b^{10}$
   (D) $a^6 + b^9$
   (E) $a^6 \cdot b^5$

5. If $\sqrt{3x} = 3.89$, then $x =$

   (A) 2.25
   (B) 5.04
   (C) 3.89
   (D) 6.51
   (E) 1.73

6. In three angles A, B, and C of a triangle, angle A exceeds angle B by 15 degrees and is twice the value of angle C. The angles A, B, and C are:

   (A) $A = 68$, $B = 63$, and $C = 49$
   (B) $A = 65$, $B = 50$, and $C = 65$
   (C) $A = 78$, $B = 63$, and $C = 39$
   (D) $A = 39$, $B = 78$, and $C = 63$
   (E) $A = 63$, $B = 39$, and $C = 78$

7. Two men start walking in opposite directions along a straight line at 4 mi/hr and 5 mi/hr, respectively. What will be the distance between them in x hours?

   (A) 9 mi
   (B) x mi
   (C) $9x$ mi
   (D) 1 mi
   (E) $20x$ mi

GO ON TO THE NEXT PAGE

MATHEMATICS LEVEL IIC TEST—Continued

8. The value of $(-5 + 2\sqrt{-4}) + (1 - \sqrt{-9}) =$

   (A) -5
   (B) -3
   (C) $-4 + i^2$
   (D) $4 - i$
   (E) $-4 + i$

9. What is the present age of John's son if 2 years ago it was one-third of John's age. Take John's present age to be x years.

   (A) $\dfrac{x-2}{3} + 2$
   (B) $\dfrac{x}{3} + 2$
   (C) $\dfrac{x}{3}$
   (D) $\dfrac{x-2}{3}$
   (E) $\dfrac{x}{3} - 2$

10. If 16.67% of a number is 12, the number is

    (A) 2
    (B) 36
    (C) 6
    (D) 72
    (E) $\dfrac{1}{6}$

11. If $f(x) = x^2 - 5x + 4$, then $f(x + a) - f(a) =$

    (A) $a^2 - 5a + 4$
    (B) $x^2 + 2ax - 5x$
    (C) $(x^2 - 5x + 4) - (a^2 - 5a + 4)$
    (D) $x^2 - 5x + 4$
    (E) 0

12. Two square tiles are placed such that they touch each other all along one side. They have areas of 36 sq.in. each. The length of a wire bounding the two tiles will be

    (A) 36
    (B) 18
    (C) 72
    (D) 48
    (E) 24

13. A cylindrical tank has a height of 12 and a radius of 3. How many cans of height 4 and radius 2 will be needed to fill the bigger tank fully?

    (A) 5
    (B) 6
    (C) 7
    (D) 8
    (E) 9

14. The domain of the function $y = \dfrac{x - 3}{(x - 2)(x + 4)}$ is

    (A) $x \neq 3$
    (B) $x \neq 2$ and $x \neq 4$
    (C) $x \geq -2$
    (D) $x \leq 4$
    (E) $-2 < x < 4$

MATHEMATICS LEVEL IIC TEST—*Continued*

15. A box is formed by cutting squares of side x from all four corners of the rectangular plate and folding along the dotted line, as shown to the right.

    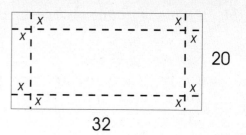

    The volume of the box is

    (A) $(32 - 2x)(20 - 2x)x$
    (B) $(32 - x)(20 - x)2x$
    (C) $640 - 4x^2$
    (D) $(32 - x)(20 - x)x$
    (E) $(32)(20)(2x)$

16. The graph of the function $y = x^2 + x - 12$ cuts the x-axis at the points

    (A) $(3, 0)$ and $(-4, 0)$
    (B) $(0, 0)$ and $(3, 0)$
    (C) $(-3, 0)$ and $(4, 0)$
    (D) $(0, 0)$ and $(-4, 0)$
    (E) $(0, 0)$ and $(12, 0)$

17. By how many degrees does the minute hand of a clock turn in 20 minutes?

    (A) $90°$
    (B) $150°$
    (C) $120°$
    (D) $180°$
    (E) $100°$

18. In the figure to the right, the coordinates of point P are

   (A) (3.46, 3.46)
   (B) (3.46, 2)
   (C) (2, 2)
   (D) (2, 3.46)
   (E) (3, 1.73)

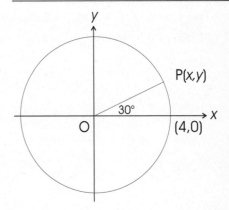

19. In triangle OAB, OA = AB, and the height of the triangle is 8.

   The slope of segment OA is

   (A) 0.375
   (B) 0.5
   (C) 2.67
   (D) 1
   (E) Cannot be determined by the given information.

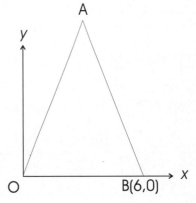

20. The graph of $y = f(x)$ is as shown above. The graph of $y = |f(x)|$ would be

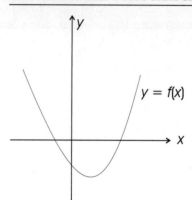

(A)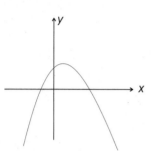

(B)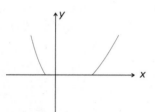

(C)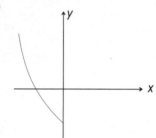

(D)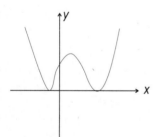

(E)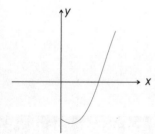

21. The equation of the line passing through the point $(0, -2)$ and perpendicular to the line $3x + 5y = 15$ is:

   (A) $3x - 5y = 10$
   (B) $3x + 5y = -10$
   (C) $5x + 3y = -6$
   (D) $5x - 3y = -6$
   (E) $5x - 3y = 6$

22. In the figure to the right, the cube has sides of length 3 each.

   The distance between points A and B is:

   (A) $3\sqrt{3}$
   (B) 3
   (C) $3 + \sqrt{3}$
   (D) 9
   (E) $\sqrt{3}$

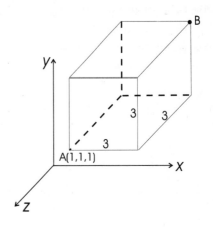

23. What is the circumference of a circle passing through all the vertices of the right triangle ABC in the figure to the right?

   (A) 78.55
   (B) 31.42
   (C) 10
   (D) 100
   (E) 24

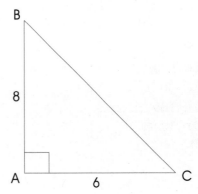

MATHEMATICS LEVEL IIC TEST—Continued

24. If $sin^2\theta - cos^2\theta = 0.5$, then $sin^4\theta - cos^4\theta =$

    (A) 0.5
    (B) 1.5
    (C) 0.25
    (D) $\sqrt{0.5}$
    (E) 0

25. A man throws a stone up an incline of gradient 1 from the base of the incline. The stone follows a path given by the equation $y^2 = 4x$. What are the coordinates of the point at which the stone hits the incline? Assume the man to be at the origin of the coordinate system.

    (A) (4, 0)
    (B) (4, 4)
    (C) (0, 4)
    (D) (3, 3)
    (E) (5, 5)

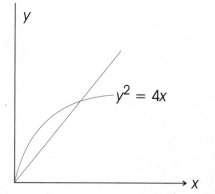

26. A triangle is cut into three regions of equal height by two lines that are parallel to the base. If the height of the triangle is h and the other dimensions are as shown to the right, what is the area of the middle region?

    (A) $0.5h$
    (B) $3h$
    (C) $18h$
    (D) $1.5h$
    (E) $9h$

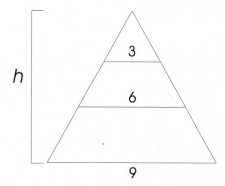

27. In the parallelogram to the right, if ∠A = 120°, then angles B, C, and D are

   (A) 60°, 120°, and 60°
   (B) 120°, 60°, and 60°
   (C) 120°, 120°, and 240°
   (D) 60°, 60°, and 120°
   (E) 180°, 120°, and 180°

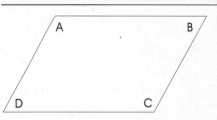

28. Consider a point P(2, 3, 4). If we are allowed to travel only parallel to the three coordinate axes, what is the distance we have to cover to get to point P from the origin?

   (A) $\sqrt{29}$
   (B) 9
   (C) 7
   (D) 5
   (E) 3

29. What is the area of the shaded region in the figure to the right?

   (A) 12.57
   (B) 8
   (C) 4.57
   (D) 7.89
   (E) 6.32

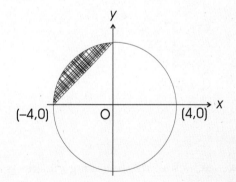

30. If an angle θ measured counter-clockwise from the positive x-axis terminates in the third quadrant, which of the following is true?

   (A) $\sin\theta$ is positive, and $\cos\theta$ is negative.
   (B) $\sin\theta$ is positive, and $\cos\theta$ is positive.
   (C) $\sin\theta$ is negative, and $\cos\theta$ is positive.
   (D) $\sin\theta$ is negative, and $\cos\theta$ is negative.
   (E) None of the above.

MATHEMATICS LEVEL IIC TEST—Continued

USE THIS SPACE FOR SCRATCH WORK

31. If $csc\theta = 1.414$, then $cos\theta =$

    (A) 0.50
    (B) 0.87
    (C) 0.71
    (D) 0.33
    (E) 0.68

32. In the figure to the right, if $cos\theta = 0.8$ and the length of segment $BC = 8$, what is the perimeter of the triangle?

    (A) 26
    (B) 16
    (C) 24
    (D) 30
    (E) 32

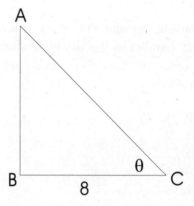

33. When the sun is 20° above the horizon, how long is the shadow that is cast by a tree 150 feet tall?

    (A) 374
    (B) 391
    (C) 412
    (D) 405
    (E) 402

34. Which of the following equations represents the curve in the figure to the right?

    (A) $y = sinx$
    (B) $y = 2sin\frac{x}{3}$
    (C) $y = 0.5sin3x$
    (D) $y = 0.5sin\frac{x}{3}$
    (E) $y = -0.5sin3x$

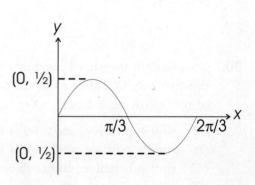

35. $\dfrac{\sqrt{1-\sin^2\theta}}{\sin\theta}$ is the same as

   (A) $\cos\theta$
   (B) $\sec\theta$
   (C) $\tan\theta$
   (D) $\csc\theta$
   (E) $\cot\theta$

36. In triangle *ABC* to the right $AB = 10$ and $BC = 8$, while $\angle A = 40°$. The value of $\angle ACB$ is

   (A) 50°
   (B) 53°
   (C) 60°
   (D) 48°
   (E) 65°

37. If triangles *ABC* and *CDE* are similar and other dimensions are as shown to the right, what is the length of segment *AE*?

   (A) 18.03
   (B) 17.05
   (C) 16.86
   (D) 19.85
   (E) 18.28

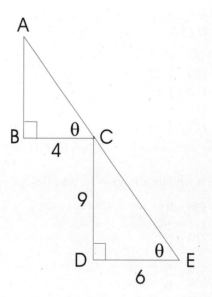

MATHEMATICS LEVEL IIC TEST—Continued

38. The resale value of a car t years after purchase is given by the function $S(t) = S_0 e^{-0.2t}$, where S_0 is the initial price of the car. If the car is purchased at $20,000, its resale value after five years will be

   (A) $6,000
   (B) $6,750
   (C) $7,000
   (D) $7,125
   (E) $7,358

39. The mean of fifteen integers is 102. On adding another integer, the mean reduces to 100. What is the new integer added?

   (A) 90
   (B) 85
   (C) 70
   (D) 65
   (E) 60

40. If $(3.5)^x = (4.2)^y$, then $\dfrac{x}{y} =$

   (A) 1.15
   (B) 1.25
   (C) 1.35
   (D) 1.45
   (E) 1.55

41. If $arcsin(sin\ x) = \dfrac{\pi}{6}$, and $0 \le x \le \dfrac{\pi}{2}$, then x could be

   (A) 0
   (B) $\dfrac{1}{2}$
   (C) $\dfrac{\pi}{6}$
   (D) $\dfrac{\pi}{3}$
   (E) $\dfrac{\pi}{2}$

42. The probability that the roll of an unbiased cubical die produces a 4 or a 6 is

   (A) $\dfrac{1}{6}$

   (B) $\dfrac{1}{3}$

   (C) $\dfrac{4}{6}$

   (D) $\dfrac{6}{6}$

   (E) $\dfrac{10}{6}$

43. In the figure to the right, the coordinates of the center, P, of the circle are

   (A) (3, 2)
   (B) (3, 8)
   (C) (3, 6)
   (D) (3, 4)
   (E) (3, 5)

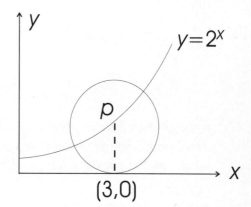

44. The value of $\ln e^3$ is

   (A) 2.718
   (B) 3.123
   (C) 3
   (D) 2.93
   (E) 3.287

MATHEMATICS LEVEL IIC TEST—Continued

45. The seventh term of an arithmetic progression is 41 and the thirteenth term is 77. What is the twentieth term of the progression?

(A) 117
(B) 118
(C) 120
(D) 121
(E) 119

46. How many possible five-digit zip codes can be formed from the set of digits {0, 1, 2 . . . 9} such that no code begins with a zero?

(A) 100,000
(B) 90,000
(C) 256
(D) 128
(E) 1,024

47. If the population of a city is increasing by 3 percent every year, and the current population is 300,000, what was the population five years ago?

(A) 258,783
(B) 250,000
(C) 265,956
(D) 311,569
(E) 303,214

48. If a, b, and x are nonzero real numbers, and if $x^4 a^3 b^5 = \dfrac{6x^3 a^5 b^2}{b^{-4}}$, then $x =$

(A) 6
(B) $6a^8 b^{11}$
(C) $6a^2 b$
(D) $6ab$
(E) $a^2 b$

49. The next three terms in the sequence of the geometric progression $-36, 12, -4, \frac{4}{3}, \ldots$ are

(A) $-\frac{4}{9}, \frac{4}{27}, -\frac{4}{81}$

(B) $\frac{4}{9}, -\frac{4}{27}, \frac{4}{81}$

(C) $-\frac{1}{3}, \frac{1}{9}, -\frac{1}{27}$

(D) $\frac{1}{3}, -\frac{1}{9}, \frac{1}{27}$

(E) $\frac{1}{3}, 0, -\frac{1}{3}$

50. What is the domain of the function defined by
$f(x) = x^2, \quad x < 3$
$ = 2x + 1, x \geq 3$

(A) $0 < x < 3$
(B) $x < 3$
(C) $x \geq 3$
(D) $x \leq 0$
(E) All real numbers.

PRACTICE TEST 1

Quick Score Answers

1. B	11. B	21. E	31. C	41. C
2. C	12. A	22. A	32. C	42. B
3. A	13. C	23. B	33. D	43. B
4. D	14. B	24. A	34. C	44. C
5. B	15. A	25. B	35. E	45. E
6. C	16. A	26. D	36. B	46. B
7. C	17. C	27. A	37. A	47. A
8. E	18. B	28. B	38. E	48. C
9. A	19. C	29. C	39. C	49. A
10. D	20. D	30. D	40. A	50. E

ANSWERS AND EXPLANATIONS

1. **The correct answer is (B).** We start with taking all terms containing x on one side and equating them with the constant terms. Thus, we have

$$1 - \frac{x}{3} = 1 - \frac{x}{2}$$

$$\Rightarrow \frac{x}{2} - \frac{x}{3} = 1 - 2$$

$$\Rightarrow \frac{3x - 2x}{6} = -1$$

$$\Rightarrow \frac{x}{6} = -1$$

$$\Rightarrow 1 + \frac{x}{6} = 0$$

2. **The correct answer is (C).** In this problem, we add the constant terms together and the terms involving the complex number together. This gives us $(2 - 5i) + (4 + 3i) = (2 + 4) + (-5i + 3i) = 6 - 2i$.

3. **The correct answer is (A).** We take the common denominator in this problem by multiplying and dividing the entire expression by (xyz). Thus we have

$$\frac{1}{x} + \frac{1}{y} + \frac{1}{z} = \frac{1}{x}\left(\frac{xyz}{xyz}\right) + \frac{1}{y}\left(\frac{xyz}{xyz}\right) + \frac{1}{z}\left(\frac{xyz}{xyz}\right)$$

$$= \frac{yz}{xyz} + \frac{xz}{xyz} + \frac{xy}{xyz}$$

$$= \frac{xy + yz + xz}{xyz}$$

Peterson's SAT II Success: Mathematics IC and IIC

ANSWERS AND EXPLANATIONS

4. **The correct answer is (D).** We use the rules of exponents in this example to simplify the original problem, as follows

   $$(a^2)^3 + \frac{b^3 \cdot b^2}{b^{-4}} = a^{2(3)} + b^{3+2-(-4)} = a^6 + b^9$$

5. **The correct answer is (B).** To find the value of x, we simplify the problem such that we can equate x to all the other constant terms.

   $$\sqrt{3x} = 3.89$$
   $$\Rightarrow \sqrt{3}(\sqrt{x}) = 3.89$$
   $$\Rightarrow \sqrt{x} = \frac{3.89}{\sqrt{3}}$$
   $$\Rightarrow x = \left(\frac{3.89}{\sqrt{3}}\right)^2$$
   $$\Rightarrow x = 5.044$$

6. **The correct answer is (C).** In this problem, we use the property of triangles, which states that the sum of internal angles in a triangle is equal to 180. Thus, $A + B + C = 180$. But $A = B + 15$, and angle C is half of A—that is, $C = \frac{A}{2}$. Thus, we have

   $$A + (A - 15) + \left(\frac{A}{2}\right) = 180$$
   $$\Rightarrow \frac{5A}{2} = 195$$
   $$\Rightarrow A = 78$$
   $$\Rightarrow B = 78 - 15 = 63, \text{ and } C = \frac{78}{2} = 39$$

7. **The correct answer is (C).** In x hours, A will have traveled $4x$ miles, and B will have traveled $5x$ miles. Since they both are traveling in opposite directions, the total distance between them will be the sum of the distance traveled in x hours, which is equal to $(4x + 5x) = 9x$ miles.

8. **The correct answer is (E).** For this problem, we recall that $i^2 = -1$. Thus, the given problem becomes

   $$(-5 + 2\sqrt{4(-1)}) + (1 - \sqrt{9(-1)}) = (-5 + 2\sqrt{4i^2}) + (1 - \sqrt{9i^2})$$
   $$= (-5 + 4i) + (1 - 3i)$$
   $$= -4 + i$$

9. **The correct answer is (A).** If John's present age is x years, his age two years ago would be $(x - 2)$ years. His son's age two years ago was $\frac{1}{3}$ of his age. So, two years ago, John's son's age was $\frac{(x-2)}{3}$. This means that John's son's present age would be $\frac{(x-2)}{3} + 2$.

10. **The correct answer is (D).** Let x be the number whose 16.67% is 12. This means $\frac{16.67x}{100} = 12$. Solving for x, we get $x = 72$.

11. **The correct answer is (B).** It is given that $f(x) = x^2 - 5x + 4$. To find $f(x + a) - f(a)$, we plug in $(x + a)$ and a for x in the original function. Thus, we have $f(x + a) - f(a) = [(x + a)^2 - 5(x + a) + 4] - [a^2 - 5(a) + 4] = x^2 + 2ax - 5x$.

12. **The correct answer is (A).** The two square tiles put together as shown in the figure below, form a rectangle with length equal to the sum of the sides of the two squared and the breadth equal to that of a side of the squares. To find the length of a wire that can bind the two square tiles, we need to find the perimeter of the rectangle.

    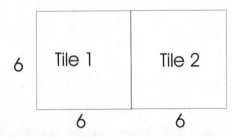

    If the squares have areas of 36 each, they have sides of length 6 each. That means that the rectangle in question has length 12 and breadth 6. The perimeter of this rectangle is $2(12 + 6) = 36$.

13. **The correct answer is (C).** For this problem, we first find the volume of the tank to be filled and the cans used to fill the tank. Then dividing the volume of the tank to be filled by that of the cans used to fill the tank, we will get the number of cans needed to fill the tank fully. The volume of a cylinder is given as $V = \pi r^2 h$, where r and h are the radius and height of the cylinder, respectively.

    Thus, the volume of the tank is $V_t = \pi(3)^2(12) = 339.29$ and the volume of the cans used to fill the tank is $V_c = \pi(2)^2(4) = 50.27$. Therefore, the number of cans needed $= \dfrac{V_t}{V_c} = \dfrac{339.29}{50.27} = 6.75$.

    Since we cannot have 6.75 cans, we round it off to the next higher figure, giving us 7 cans as the answer.

14. **The correct answer is (B).** To find the domain of the function, $f(x) = \dfrac{x - 3}{(x - 2)(x + 4)}$, we observe that the function is not defined only if the denominator becomes equal to zero. That will happen only if $(x - 2) = 0$ or $(x + 4) = 0$. This gives us $x = 2$ or $x = -4$ as values not allowed. The function is defined for all other values of x.

15. **The correct answer is (A).** After cutting squares of side x from each corner and folding along the dotted lines, we end up with a box whose height is x and whose base has dimensions $(32 - 2x)$ and $(20 - 2x)$, since $2x$ is cut off from the length and width, respectively, of the rectangular sheet. This makes the dimensions of the box $(32 - 2x)(20 - 2x)(x)$.

ANSWERS AND EXPLANATIONS

16. **The correct answer is (A).** To find the point at which a curve cuts the x-axis, we use the fact that any point on the x-axis has a y-coordinate equal to zero. Thus, plugging in $y = 0$ in the equation of a given curve and solving for x, we can obtain the x-coordinate of the point at which the curve cuts the x-axis. Thus, we have

    $0 = x^2 + x - 12$

    $\Rightarrow (x + 4)(x - 3) = 0$

    $\Rightarrow x = -4$ or $x = 3$

    Hence, the two points at which the given function cuts the x-axis are $(0, -4)$ and $(0, 3)$.

17. **The correct answer is (C).** In 60 minutes, the minute hand of a clock does one complete revolution or rotates by 360°. Therefore, in 20 minutes, it will move by $\left(\dfrac{20}{60}\right)(360) = 120°$.

18. **The correct answer is (B).** For this problem, we just use the formulae of "polar coordinates" of a circle. For any point $P(x, y)$ on the circle, we have $x = r(\cos\theta)$ and $y = r(\sin\theta)$, where r = the radius of the circle and θ = the angle made by P with the positive direction of the x-axis. For $r = 4$ and $\theta = 30°$, we have $x = 4(\cos 30) = 3.46$ and $y = 4(\sin 30) = 2$. Therefore, the coordinates of point P are $(3.46, 2)$.

19. **The correct answer is (C).** In triangle OAB, since $OA = AB$, we know that the perpendicular bisector of segment OB will pass through point A. This gives us the x-coordinate of point A as 3. The y-coordinate of point A has to be 8, since that is also the height of the triangle. Now, we need to find the slope of the line passing through $A(3, 8)$ and $O(0, 0)$, which is $\dfrac{8 - 0}{3 - 0} = 2.67$.

20. **The correct answer is (D).** The absolute value of a function is the set of all absolute values (without the sign) that the function takes. That means if $f(x) = -k$ then $|f(x)| = k$. For the given curve, all the negative values (part of the curve below the x-axis) become positive, resulting in the curve in choice (D).

21. **The correct answer is (E).** The slopes, $m1$ and $m2$, of two lines perpendicular to one another are related such that $(m1)(m2) = -1$. The slope of line $3x + 5y = 15$ is $m1 = \dfrac{-3}{5}$. Therefore, the slope of a line that is perpendicular to it will be $m2 = \dfrac{-1}{\frac{-3}{5}} = \dfrac{5}{3}$. The equation of this line, if it passes through the point $(0, -2)$, will be $y = \dfrac{5}{3}x - 2$ (using the form $y = mx + b$).

22. **The correct answer is (A).** For the given cube of side 3 units, the coordinates of point B would be $(4, 4, -2)$. The distance between two points $(x1, y1, z1)$ and $(x2, y2, z2)$ is given as $D = \sqrt{(x2 - x1)^2 + (y2 - y1)^2 + (z2 - z1)^2}$. Using this formula, the distance between points $A(1, 1, 1)$ and $B(4, 4, -2)$ will be $D = \sqrt{(4 - 1)^2 + (4 - 1)^2 + (-2 - 1)^2} = \sqrt{9 + 9 + 9} = 3\sqrt{3}$.

23. **The correct answer is (B).** For this problem, we make use of the property of circles, which states that any point on the circumference of a circle forms a right-angled triangle with a diameter of the circle such that the diameter is the hypotenuse of the right triangle. Thus, for the given triangle, the hypotenuse, AC, forms a diameter of the circle that passes through the points A, B, and C. The length of side AC of the triangle is determined by the Pythagorean theorem as $= 10$, which means that the diameter of the circle $= 10$. The circumference of this circle $= \pi(d) = \pi(10) = 31.42$.

24. **The correct answer is (A).** We have been given that $(sin^2\theta - cos^2\theta) = 0.5$. Now, $(sin^4\theta - cos^4\theta) = (sin^2\theta + cos^2\theta)(sin^2\theta - (cos^2\theta) = (1)(0.5) = 0.5$.

25. **The correct answer is (B).** The incline has a gradient (or slope) of 1, which means that the equation of the line is $y = x$. Finding the coordinates of the point where the stone hits the incline is the same as finding the point of intersection of the two curves $y = x$ and $y^2 = 4x$.

    Solving the two equations simultaneously, we have $x = 0$ or $x = 4$. Since the value $x = 0$ does not apply in this case, we have $x = 4$, which gives us $y = 4$. Thus, the point at which the stone hits the incline is $(4, 4)$.

26. **The correct answer is (D).** The region of concern forms a trapezium of height $\left(\dfrac{h}{3}\right)$ and with bases 3 and 6. The area of this trapezium is $\dfrac{3 + 6}{2}\left(\dfrac{h}{3}\right) = 1.5h$.

27. **The correct answer is (A).** Using the property of internal angles of a parallelogram, we have the sum of adjacent angles $= 180°$ and the opposite angles equal to each other. This gives us $C = A$, $(A + B) = 180$, and $B = D$.

    Using the above relations, we have $B = 60°$, $C = 120°$, and $D = 60°$.

28. **The correct answer is (B).** Since we are allowed to travel only parallel (or along) the three coordinate axes to reach to a point $(2, 3, 4)$ in space, we have to travel 2 units along the x-axis, then 3 units parallel to the y-axis, and finally 4 units parallel the z-axis. Thus, we will have to travel a total of $(2 + 3 + 4) = 9$ units.

ANSWERS AND EXPLANATIONS

29. **The correct answer is (C).** To get the area of the shaded region, we first obtain the area of the part of the circle in that quadrant and subtract from it the area of the triangle *OAB*. The area of the part of the circle in the second quadrant

    $= \frac{1}{4}$ (area of the circle)

    $= \frac{1}{4}(\pi)(4)^2$

    $= 12.566$

    The area of the triangle $OAB = \frac{1}{2}(4)(4) = 8$. Therefore, the net area $= (12.566) - 8 = 4.566$

30. **The correct answer is (D).** In the third quadrant, the tangent of an angle is positive, while the sine and cosine of the angle are negative.

31. **The correct answer is (C).** If $csc\theta = 1.414$, then $sin\theta = \frac{1}{csc\theta} = \frac{1}{1.414}$. Using the trigonometric identity, $sin^2\theta + cos^2\theta = 1$, we have $cos\theta = \sqrt{1 - sin^2\theta} = \sqrt{1 - \left(\frac{1}{1.414}\right)^2} = 0.71$.

32. **The correct answer is (C).** By definition, $cos\theta = \frac{BC}{AC}$. If $cos\theta = 0.8$ and $l(BC) = 8$, then $0.8 = \frac{8}{AC} \Rightarrow l(AC) = 10$.

    To find the length of *AB*, we use the Pythagorean theorem, giving us $l(AB) = \sqrt{(10)^2 - (8)^2} = 6$. Therefore, the perimeter of the triangle is $= 8 + 10 + 6 = 24$ units.

33. **The correct answer is (D).** We refer to the following figure to solve this problem. If *x* is the length of the shadow cast by the tree, then $tan\ 20 = \frac{150}{x} \Rightarrow x = \frac{150}{tan\ 20} = 412$ ft.

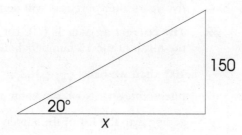

34. **The correct answer is (C).** From the figure in the problem, we observe that the given curve has a maximum value of $\frac{1}{2}$ and starts to repeat itself after $x = \frac{2\pi}{3}$. This means that it has an amplitude of $\frac{1}{2}$ and period of $x = \frac{1}{3}(2\pi)$. This gives the equation of the sine curve as $y = \frac{1}{2}\sin\left(\frac{x}{\frac{1}{3}}\right) = \frac{1}{2}\sin(3x)$.

35. **The correct answer is (E).** $\frac{\sqrt{1-\sin^2\theta}}{\sin\theta} = \frac{\sqrt{\cos^2\theta}}{\sin\theta} = \frac{\cos\theta}{\sin\theta} = \cot\theta$.

36. **The correct answer is (B).** In this problem, we make use of the Sine Rule of Triangles, which is as follows: $\frac{\sin A}{l(BC)} = \frac{\sin B}{l(AC)} = \frac{\sin C}{l(AB)}$.

    Thus, $\frac{\sin 40}{8} = \frac{\sin C}{10} \Rightarrow \sin C = \frac{10(\sin 40)}{8} = 0.8 \Rightarrow C = \sin^{-1}(0.8) = 53°$.
    Therefore, $\angle ACB = 53°$.

37. **The correct answer is (A).** In this problem, triangles ABC and CDE are similar. This gives us:

    $\frac{l(AB)}{l(CD)} = \frac{l(BC)}{l(DE)} = \frac{l(AC)}{l(CE)} \Rightarrow \frac{l(AB)}{9} = \frac{4}{6} \Rightarrow l(AB) = 6$.

    Now, $AE = AC + CE$. To get the lengths of AC and CE, we use the Pythagorean theorem. Thus, $l(AC) = \sqrt{4^2 + 6^2} = \sqrt{52} = 7.21$ and $l(CE) = \sqrt{9^2 + 6^2} = \sqrt{117} = 10.82$. Therefore, $l(AE) = 10.82 + 7.21 = 18.03$.

38. **The correct answer is (E).** The function $S(t) = S_0 e^{-0.2t}$ represents the sale price of the car t years after purchase, with S_0 representing the initial price of the car. To find the resale value of the car in five years, given that the car was initially priced at $20,000, we plug in the values of t and S_0 and obtain the corresponding value of $S(t)$. Thus, we have $S(5) = (20,000)e^{-0.2(5)} = 7,358$. Therefore, the resale value of the car five years after purchase will be $7,358.

39. **The correct answer is (C).** Let the new added integer be x. Now, if the mean of the 15 numbers before we added the new integer was 102, then we have: $\frac{S}{15} = 102 \Rightarrow S = 1530$, where S is the sum of the fifteen integers. Now, we know that the mean changes to 100 after adding x to the list of the fifteen integers. This gives us $\frac{S+x}{16} = 100 \Rightarrow (S+x) = 1,600 \Rightarrow x = 1,600 - S = 70$.

ANSWERS AND EXPLANATIONS

40. **The correct answer is (A).** $(3.5)^x = (4.2)^y$. Taking natural logs on both sides and using the property $\ln x^k = k\ln x$, we have

    $\ln(3.5)^x = \ln(4.2)^y$

    $\Rightarrow x \ln(3.5) = y \ln(4.2)$

    $\Rightarrow \dfrac{x}{y} = \dfrac{\ln(4.2)}{\ln(3.5)} = 1.15.$

41. **The correct answer is (C).** By definition, $arcsin(sin\theta) = \theta$. Therefore, if $arcsin(sin\theta) = \dfrac{\pi}{6}$, then $\theta = \dfrac{\pi}{6}$.

42. **The correct answer is (B).** A roll of an unbiased die has equal probability to produce any of the six possible outcomes (1, 2, 3, 4, 5, or 6). Thus, the probability that a 4 or a 6 turns up will be $\dfrac{2}{6}$ or $\dfrac{1}{3}$.

43. **The correct answer is (B).** In this problem, we see that the center of the circle, $P(x, y)$, lies on the curve $y = 2^x$. Hence the point P must satisfy the equation of the curve. Now, the x-coordinate of point P is 3. To obtain the y-coordinate, we plug in 3 for x in the equation of the curve. Thus, $y = 2^3 = 8$. Therefore, the point P has coordinates (3, 8).

44. **The correct answer is (C)** Here again, using the property $\ln x^k = k \ln x$ of log functions and keeping in mind that $\ln e = 1$, we have $\ln e^3 = 3\ln e = 3(1) = 3$.

45. **The correct answer is (E).** Let a be the first term and x be the common difference of the arithmetic progression. Then, by definition, the seventh term will be $(a + 6x)$ and the thirteenth term will be $(a + 12x)$. We have been given that the seventh term is 41 and the thirteenth term is 77. Thus, we have

    $a + 6x = 41$, and
    $a + 12x = 77$.

    Solving the above two equations simultaneously, we get $a = 5$ and $x = 6$. Therefore, the twentieth term will be $a + 19x = 5 + 19(6) = 119$.

46. **The correct answer is (B).** Since no zip codes can begin with a zero, we have nine digits to choose from for the first of the five-digit code. Also, since there is no restriction on repeating the digits, the remaining four digits of the five-digit code can each take 10 values. Therefore, we have a total of $9(10)(10)(10)(10) = 9(10)^4 = 90{,}000$ options.

47. **The correct answer is (A).** This problem can be solved using the formula of exponential growth, which is $A = P(1 + r)^t$, where A = the amount present finally, P = the initial amount, r = the rate of growth, and t = the total time. Thus, for our problem, we have $300{,}000 = P(1 + 0.03)^5 = \Rightarrow P = 258{,}783$.

PRACTICE TEST 1

48. **The correct answer is (C).** In this problem, we first group all terms in x, a, and b together and then use the properties $u^p \cdot u^q = u^{p+q}$ and $\dfrac{u^p}{u^q} = u^{p-q}$ to simplify the problem. Thus, we have

$$x^4 a^3 b^5 = \frac{6x^3 a^5 b^2}{b^{-4}}$$

$$\Rightarrow \frac{x^4}{x^3} = \frac{6a^5 b^2}{b^{-4} a^3 b^5}$$

$$\Rightarrow x^{4-3} = 6a^{5-3} b^{2-5-(-4)}$$

$$\Rightarrow x = 6a^2 b.$$

49. **The correct answer is (A).** The given geometric progression is: $-36, 12, -4, \dfrac{4}{3}, \ldots$ Here, $\dfrac{12}{-36} = \dfrac{-4}{12} = \dfrac{-1}{3}$. Thus, the constant multiple of the progression is $\dfrac{-1}{3}$. Therefore, the next three terms in the progression will be

$$\frac{4}{3}\left(\frac{-1}{3}\right), \frac{4}{3}\left(\frac{-1}{3}\right)^2, \frac{4}{3}\left(\frac{-1}{3}\right)^3 = \frac{-4}{9}, \frac{4}{27}, \frac{-4}{81}.$$

50. **The correct answer is (E).** We observe that the function

$$f(x) = x^2, \quad x < 3$$
$$= 2x + 1, \quad x \geq 3$$

is defined for all possible values of x.

Practice Test 2

MATHEMATICS LEVEL IIC TEST

PRACTICE TEST 2

MATHEMATICS LEVEL IIC TEST

While you have taken many standardized tests and know to blacken completely the ovals on the answer sheets and to erase completely any errors, the instructions for the SAT II Mathematics IIC exam differ in three important ways from the directions for other standardized tests you have taken. You need to indicate on the answer key which test you are taking.

The instructions on the answer sheet will tell you to fill out the top portion of the answer sheet exactly as shown.

1. Print MATHEMATICS LEVEL IIC on the line to the right under the words *Subject Test (print)*.

2. In the shaded box labeled *Test Code* fill in four ovals:

    —Fill in oval 5 in the row labeled V.

    —Fill in oval 3 in the row labeled W.

    —Fill in oval 5 in the row labeled X.

    —Fill in oval E in the row labeled Y.

    —Leave the ovals in row Q blank.

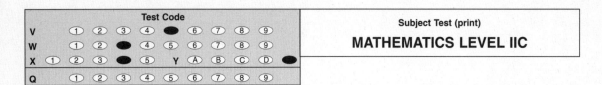

When everyone has completed filling in this portion of the answer sheet, the supervisor will tell you to turn the page and begin the Mathematics Level IIC examination. The answer sheet has 100 numbered ovals on the sheet, but there are only 50 multiple-choice questions in the test, so be sure to use only ovals 1 to 50 to record your answers.

MATHEMATICS LEVEL IIC TEST

REFERENCE INFORMATION

The following information is for your reference in answering some of the questions in this test.

Volume of a right circular cone with radius r and height h:
$$V = \frac{1}{3}\pi r^2 h$$

Lateral area of a right circular cone with circumference of the base c and slant height l:
$$S = \frac{1}{2}cl$$

Surface area of a sphere with radius r:
$$S = 4\pi r^2$$

Volume of a pyramid with base area B and height h:
$$V = \frac{1}{3}Bh$$

For each of the following problems, identify the BEST answer of the choices given. If the exact numerical value is not one of the choices, select the answer that is closest to this value. Then fill in the corresponding oval on the answer sheet.

NOTES

1. You will need a calculator to answer some (but not all) of the questions. You must decide whether or not to use a calculator for each question. You must use at least a scientific calculator; you are permitted to use graphing or programmable calculators.

2. Degree measure is the only angle measure used on this test. Be sure your calculator is set in degree mode.

3. The figures that accompany questions on the test are designed to give you information that is useful in solving problems. They are drawn as accurately as possible EXCEPT when stated that a figure is not drawn to scale. Unless otherwise indicated, all figures lie in planes.

4. The domain of any function f is assumed to be the set of all real numbers x for which $f(x)$ is a real number, unless otherwise specified.

MATHEMATICS LEVEL IC TEST

1. Which number is not in the domain of $y = \dfrac{x+2}{x+3}$?

    (A) 2
    (B) −2
    (C) −3
    (D) 3
    (E) 0

2. What trigonometric function(s) is (are) positive in the third quadrant?

    (A) *sin x*
    (B) *cos x*
    (C) *sin x* and *cos x*
    (D) *tan x* and *cot x*
    (E) *sin x* and *csc x*

3. If the vertex of a function *f(x)* is at (1, 1), where is the vertex of the function *f(x* − 2) + 1?

    (A) (2, 3)
    (B) (3, 2)
    (C) (2, −1)
    (D) (−1, 0)
    (E) (−1, 2)

4. You have 7 marbles—2 black, 3 white, and 2 red, but otherwise not distinguishable. How many different ways can the 7 marbles be ordered?

    (A) 5,040
    (B) 84
    (C) 140
    (D) 210
    (E) 280

5. A church bell chimes 8 times at 8:00. Eight seconds elapse between the first chime and the last chime. How many seconds elapse between the first and last chimes at 12:00? (Assume each actual chime takes no time at all.)

(A) 12
(B) 12.5
(C) 12.57
(D) 13
(E) 11

6. Given the figure to the right, find y.

(A) 4
(B) 2
(C) 3
(D) 6
(E) 8

7. If
$f = \{(0,-3),(-1,-2),(2,-1)\}$ and
$g = \{(-1,2),(-2,1),(-3,3)\}$, then
$g \circ f(-1)$ equals

(A) -1
(B) 1
(C) 2
(D) -2
(E) Not defined

MATHEMATICS LEVEL IC TEST—Continued

8. What is the period of $y = 3\cos(2x) + 4$?

   (A) $\dfrac{2\pi}{3}$
   (B) $\dfrac{\pi}{2}$
   (C) 4π
   (D) π
   (E) 3π

9. Solve $2\sin x + 3 = 4$ for x in the interval $(0, 360)$. (Remember, x is measured in degrees.)

   (A) 60
   (B) 30 and 150
   (C) 30
   (D) 60 and 120
   (E) 150

10. Find the volume of the solid that is a cylinder with a section cut out of it.

    (A) 28π cubed meters
    (B) 12 cubed meters
    (C) 12π cubed meters
    (D) 24π cubed meters
    (E) $\dfrac{56\pi}{3}$ cubed meters

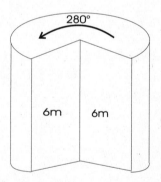

11. If $x + 2 < 0$, then $|x + 2| =$

    (A) $x + 2$
    (B) $x - 2$
    (C) $-x + 2$
    (D) $-x - 2$
    (E) x

MATHEMATICS LEVEL IC TEST—Continued

12. What are the asymptotes of the hyperbola $4x^2 - 9y^2 = 36$?

    (A) $y = 9x$ and $y = -9x$
    (B) $y = \dfrac{3x}{2}$ and $y = \dfrac{-3x}{2}$
    (C) $y = 2x$ and $y = -2x$
    (D) $y = \dfrac{2x}{3}$ and $y = \dfrac{-2x}{3}$
    (E) $y = 4x$ and $y = -4x$

13. There are six movie stars who pass through towns A and B in a certain state. Of these, $\dfrac{1}{2}$ stop at A, $\dfrac{1}{3}$ stop at B, and $\dfrac{1}{6}$ stop at both A and B. How many movie stars don't stop at either town?

    (A) 0
    (B) 1
    (C) 2
    (D) 3
    (E) Cannot be determined

14. Given that in the figure to the right, s is parallel to t and v is parallel to w, find the angle measure of angle 1.

    (A) 65
    (B) 55
    (C) 45
    (D) 125
    (E) 115

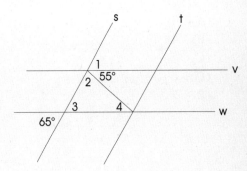

MATHEMATICS LEVEL IC TEST—Continued

15. If $2x + 3$ is a divisor of $2x^3 + 7x^2 + 8x + c$ with a remainder of 0, c is

    (A) -33
    (B) -3
    (C) 33
    (D) $-\dfrac{3}{2}$
    (E) 3

16. If we restrict the domain of the $f(x) = x^2 + 3$ to $(-2, 1)$, then the range of $f(x)$ is

    (A) $(3, 7)$
    (B) All positive real numbers
    (C) $(3, 4)$
    (D) $(4, 7)$
    (E) $(0, 7)$

17. Two distinct lines can intersect one time at most and three distinct lines can intersect three times at most. What is the greatest number of times that four distinct lines can intersect?

    (A) 3
    (B) 4
    (C) 5
    (D) 6
    (E) 7

MATHEMATICS LEVEL IC TEST—Continued

USE THIS SPACE FOR SCRATCH WORK

18. If a square field is completely enclosed by x feet of fencing, then the area of the field as a function of x equals

   (A) x^2
   (B) $\dfrac{x^2}{4}$
   (C) $4x^2$
   (D) $\dfrac{x^2}{16}$
   (E) $16x^2$

19. If $f(x) = x^2 + x$, then find the number(s) so that $f(a) = 6$.

   (A) 3
   (B) -2
   (C) 3 and -2
   (D) -3 and 2
   (E) -3

20. What is the exact value of $\tan\left(\text{Arccos }\dfrac{3}{4}\right)$?

   (A) $\dfrac{-3\sqrt{7}}{7}$
   (B) $\dfrac{3\sqrt{7}}{7}$
   (C) $\dfrac{-\sqrt{7}}{3}$
   (D) $\dfrac{3}{4}$
   (E) $\dfrac{\sqrt{7}}{3}$

GO ON TO THE NEXT PAGE

MATHEMATICS LEVEL IC TEST—Continued

21. Given the graph to the right, find the amplitude.

    (A) .5
    (B) π
    (C) $\dfrac{\pi}{2}$
    (D) 1.5
    (E) 2

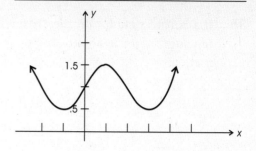

22. Solve the equation $\log_2 (x - 3) + \log_2 (x - 2) = 1$ for x.

    (A) 3
    (B) 2 and 3
    (C) 1 and 4
    (D) 4
    (E) 1

23. In a dark room where colors are not distinguishable, how many towels must a person take from a basket containing 10 blue towels, 8 black towels and 6 green towels—to be assured of having two towels that match in color?

    (A) 1
    (B) 2
    (C) 3
    (D) 4
    (E) 5

24. Find the next number in the sequence 1, 7, 19, 37, 61, ?, . . .

    (A) 85
    (B) 78
    (C) 91
    (D) 95
    (E) 90

MATHEMATICS LEVEL IC TEST—Continued

USE THIS SPACE FOR SCRATCH WORK

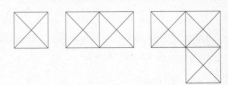

25. Given the square with two diagonals in the figure to the right, there are a maximum of eight total triangles that can be formed. With two squares with diagonals adjoined on one side, there are a maximum of eighteen total triangles that can be formed. What is the maximum number of triangles that can be formed?

   (A) 26
   (B) 29
   (C) 28
   (D) 30
   (E) 27

26. The function $f(x) = x^2$ is an example of a(n)

   (A) even function
   (B) polynomial
   (C) quadratic function
   (D) all of these choices
   (E) only choices (B) and (C)

27. Simplify the expression $tan(-\eta)cos(-\eta)$ in terms of a positive *angle* η, $sin\ \eta$ and $cos\ \eta$.

   (A) $sin\ \eta$
   (B) $cos\ \eta$
   (C) $-sin\ \eta$
   (D) $-cos\ \eta$
   (E) $tan\ \eta$

28. To draw the graph of the inverse of a function $f(x)$, one must mirror the graph of $f(x)$ about the

   (A) x-axis
   (B) y-axis
   (C) line $y = -x$
   (D) line $y = x$
   (E) lines $y = x$ and $y = -x$

GO ON TO THE NEXT PAGE

MATHEMATICS LEVEL IC TEST—Continued

29. In numbering the pages of a book, beginning with page 1, 3,457 digits are required. What is the number of pages in the book?

   (A) 1,003
   (B) 3,457
   (C) 1,141
   (D) 1,140
   (E) 1,138

30. $\left(1 - \dfrac{x}{y}\right) \div \left(1 - \dfrac{x^2}{y}\right)$ simplifies to

   (A) $y - x$
   (B) $y + x$
   (C) $(y + x)^{-1}$
   (D) $(y - x)^{-1}$
   (E) $(y - x)(y + x)^{-1}$

31. Find c in the figure to the right.

   (A) 2.71
   (B) 8.00
   (C) 2.83
   (D) 5.03
   (E) 3.71

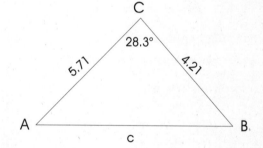

32. Find x in the figure to the right. (DC ∥ PQ)

   (A) 2
   (B) 6
   (C) 4
   (D) 8
   (E) 10

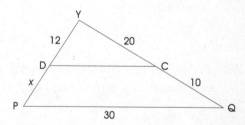

MATHEMATICS LEVEL IC TEST—Continued

USE THIS SPACE FOR SCRATCH WORK

33. Find all the values of x that satisfy $(2x + 1)^{x^2 - 4} = 1$.

    (A) 2 and 0
    (B) −2 and 1
    (C) 1 and 2
    (D) 2 and −2
    (E) None of the above

34. How many subsets are there from a set of m elements?

    (A) m
    (B) 2^m
    (C) m^2
    (D) $m!$
    (E) $m(m - 1)$

35. Find the volume of the prism to the right.

    (A) 380 cubed centimeters
    (B) 480 cubed centimeters
    (C) 600 cubed centimeters
    (D) 720 cubed centimeters
    (E) 1,200 cubed centimeters

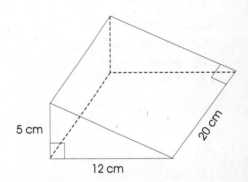

GO ON TO THE NEXT PAGE

MATHEMATICS LEVEL IC TEST—Continued

36. The axis of symmetry for $f(x) = x^2 - 2x + 3$ is $x =$

(A) 2
(B) 1
(C) -2
(D) -1
(E) $\frac{1}{2}$

37. The figure to the right shows three of the faces of a cube. If the six faces of the cube are numbered consecutively, what are the possible values for the product of all six faces?

I. 5,040
II. 20,160
III. 720

(A) I only
(B) II only
(C) III only
(D) I and II
(E) I and III

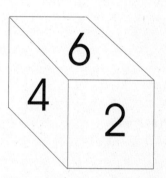

38. If $ax^n - bx = 0$ and x is nonzero, then b equals

(A) ax^{n-1}
(B) ax^{2n}
(C) a^n
(D) na
(E) ax^{n+1}

39. If $f(x) = 2\sin 3x$ and $f\left(\dfrac{\pi}{2}\right) = b$, then find b.

(A) 1
(B) −1
(C) 2
(D) −2
(E) 0

40. Simplify $\sin^2 x + \dfrac{(\cos 2x)}{2} + \dfrac{1}{2}$

(A) $\sin 2x$
(B) 1
(C) $\sin^2 x + \sin\left(\dfrac{x}{2}\right)$
(D) $\sin^2 x + \cos\left(\dfrac{x}{2}\right)$
(E) −1

41. If the equation $\dfrac{x^2}{4-C} - \dfrac{y^2}{C} = 1$, what type of curve is represented if $C < 0$?

(A) circle
(B) ellipse
(C) hyperbola
(D) parabola
(E) line

42. What is the units' digit of 3^{30}?

(A) 9
(B) 3
(C) 7
(D) 1
(E) 2

MATHEMATICS LEVEL IC TEST—Continued

USE THIS SPACE FOR SCRATCH WORK

43. A rectangle has a height of 8 units and a width of 6 units. A second rectangle with a height of 4 units and a width of 3 units overlaps the first rectangle as shown in the figure to the right. What is the difference in area between the two nonoverlapping regions of the two rectangles?

   (A) 36 square units
   (B) 12 square units
   (C) 24 square units
   (D) 30 square units
   (E) 18 square units

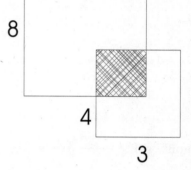

44. If $x^{-2} = 36$, then x equals

   (A) ± 6
   (B) $\pm \frac{1}{6}$
   (C) $\pm 6i$
   (D) ± 18
   (E) $\pm \frac{i}{6}$

45. A pit filled with balls for kids at an amusement park is 10 feet long and 2 feet deep at one end. If it is 4 feet deep at the other end, find the total distance along the bottom in feet.

   (A) 10.77
   (B) 10
   (C) 4.47
   (D) 11
   (E) 10.20

MATHEMATICS LEVEL IC TEST—Continued

46. What must be the value of k if the lines $3x - y = 9$ and $kx + 3y = 5$ are to be perpendicular?

    (A) 1
    (B) 3
    (C) 9
    (D) −1
    (E) −9

47. Find the sum of the first 40 odd integers.

    (A) 1,000
    (B) 1,600
    (C) 1,200
    (D) 660
    (E) None of the above

48. If $f(x) = 2x + 3$ and $f(g(1)) = 5$, which of the following functions could be $g(x)$?

    (A) $x + 1$
    (B) $2x + 1$
    (C) $2x - 1$
    (D) $x - 1$
    (E) $3x - 4$

49. If Bill can clean the garage in 4 hours and John can clean the same garage in 3 hours, how many hours will it take them to clean the garage if they work together?

    (A) 3
    (B) 3.5
    (C) 1.71
    (D) .58
    (E) 2

MATHEMATICS LEVEL IC TEST—Continued

50. What is the perimeter of the parallelogram ACDE in the figure to the right?

    (A) 20
    (B) 23
    (C) 23.31
    (D) 33.94
    (E) 18

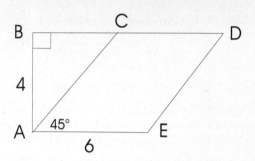

ANSWERS AND EXPLANATIONS

Quick Score Answers

1. C	11. D	21. A	31. C	41. B
2. D	12. D	22. D	32. B	42. A
3. B	13. C	23. D	33. A	43. A
4. D	14. A	24. C	34. B	44. B
5. C	15. E	25. B	35. C	45. E
6. B	16. A	26. D	36. B	46. A
7. B	17. D	27. C	37. E	47. B
8. D	18. D	28. D	38. A	48. C
9. B	19. D	29. C	39. D	49. C
10. E	20. E	30. C	40. B	50. C

ANSWERS AND EXPLANATIONS

1. **The correct answer is (C).** The function y is defined for all numbers except where $x + 3 = 0$, which is when $x = -3$.

2. **The correct answer is (D).** We know $sin\ x < 0$ in the third quadrant and so that eliminates choices (A), (C), and (E). Also, $cos\ x < 0$ in the third quadrant, which eliminates choice (B).

3. **The correct answer is (B).** The graph of $f(x - 2) + 1$ is just the graph of $f(x)$ moved to the right 2 units and up 1 unit. So, the vertex of $f(x - 2) + 1$ is just the vertex (1, 1) moved to the right 2 units and up 1 unit. Hence, the vertex is (3, 2).

4. **The correct answer is (D).** There are seven different places where a marble can be placed, so there are a total of 7! orderings. However, some orderings look exactly like each other. For example, if we denote the white marbles W1, W2, and W3; the black marbles as B1 and B2; and the red marbles as R1 and R2, we see the orderings

   R1, R2, B1, B2, W1, W2, W3 is the same as
   R2, R1, B1, B2, W1, W2, W3, but it is not the same as
   R2, R1, W1, B1, B2, W2, W3.

   We do not want to include the permutations of just the black marbles, just the red marbles, and just the white marbles, which correspond to dividing by 2!(2!)(3!) = 24. So the answer is $\frac{7!}{24} = 210$.

5. **The correct answer is (C).** There are 8 chimes in 8 seconds, and we assume that no actual time is taken up for the chime itself. Once the first chimes, there is a space of time before the next chimes and so on until the last chimes and then the 8 seconds is up. There are 7 spaces of time with $\frac{8}{7}$ seconds elapsing between chimes. For 12 chimes, there are 11 spaces of time with $11 \times \left(\frac{8}{7}\right)$ seconds of time or 12.57 seconds.

Peterson's: www.petersons.com

PRACTICE TEST 2

6. **The correct answer is (B).** Each side of the triangle is tangent to the circle. So, the length of line segments from a vertex to the point of tangency is equal. That is, $x + 4 = 6$ or $x = 2$ and $x + y = 2 + y = 4$, which says that $y = 2$.

7. **The correct answer is (B).** $(g \circ f)(-1) = g(f(-1)) = g(-2) = 1$.

8. **The correct answer is (D).** The period is $p = \dfrac{2\pi}{b} = \dfrac{2\pi}{2} = \pi$.

9. **The correct answer is (B).** We have $2\sin x + 3 = 4$, which implies $\sin x = \dfrac{1}{2}$. That says that $x = 30$ or $x = 150$.

10. **The correct answer is (E).** The area of a full circle with radius 2 is 4π, so the area of the top is $\dfrac{280}{360} \cdot 4\pi = \dfrac{28\pi}{9}$ and, thus, the volume is $V = 6 \times \left(\dfrac{28\pi}{9}\right) = \dfrac{56\pi}{3}$.

11. **The correct answer is (D).** If $x + 2 < 0$, then $x + 2$ is negative, and, so by the definition of absolute value, $|x + 2| = -(x + 2) = -x - 2$.

12. **The correct answer is (D).** We see that the given equation is equivalent to $\dfrac{x^2}{9} - \dfrac{y^2}{4} = 1$. The asymptotes are $y^2 = \dfrac{4x^2}{9}$ or $y = \pm \dfrac{2x}{3}$. To see this, we know that $a = 3$, which says the vertices are $(\pm 3, 0)$. Also, since $b = 2$, the other vertices are $(0, \pm 2)$. Since the center of the hyperbola is $(0, 0)$, the asymptotes go through the origin, and when we calculate the lines that pass through the origin and the points $(3, 2)$ and $(3, -2)$ we get the desired answers.

13. **The correct answer is (C).** To see this the best, make a Venn diagram and label how many visited the intersection of Town A and Town B and then from that how many visited Town A, and how many visited Town B. This will give you how many visited neither Town A nor Town B. We know that one-half visited Town A which corresponds to 3. Similarly, one-third visited Town B, which corresponds to 2, and finally, one-sixth visited both Town A and Town B, which corresponds to 1. This yields that 2 movie stars visited neither Town A nor Town B.

14. **The correct answer is (A).** By vertical angles, we know that angle 3 is 65, and by alternate angles, angle 1 is 65.

15. **The correct answer is (E).** If $2x + 3$ is a divisor of $2x^3 + 7x^3 + 8x + c$, then $-\dfrac{3}{2}$ is a root. Use synthetic division or use long division to get that $c = 3$ because the remainder is 0.

16. **The correct answer is (A).** The range of $f(x)$ with no restrictions is all real numbers bigger than or equal to 3. If we restrict to the domain of $(-2, 1)$ then the range is $(3, 7)$ since $f(-2) = (-2)^2 + 3 = 7$.

17. **The correct answer is (D).** First draw three lines that cross at three points (there will be a triangle formed), then draw one more line that crosses all of the other three lines at distinct points. The result is that three lines intersect in six points at most.

ANSWERS AND EXPLANATIONS

18. **The correct answer is (D).** We have the perimeter to be x, and since the perimeter is $4s$ where s is the length of a side, then $s = \frac{x}{4}$. So, the area is $A = s^2 = \left(\frac{x}{4}\right)^2 = \frac{x^2}{16}$.

19. **The correct answer is (D).** $f(a) = a^2 + a = 6$, which implies that $a^2 + a - 6 = 0$. By factoring we get $(a + 3)(a - 2) = 0$, and so $a = 2$ or $a = -3$.

20. **The correct answer is (E).** Let $\theta = Arccos\left(\frac{3}{4}\right)$. This says that $\cos\theta = \frac{3}{4}$. Label a triangle with angle θ and $\cos\theta = \frac{3}{4}$. Find the other side, which happens to be $\sqrt{7}$, and then we see that $\tan\theta$ is the desired result.

21. **The correct answer is (A).** The maximum value is 1.5, and the minimum value of y value is .5, so the amplitude is $\frac{(1.5 - .5)}{2} = \frac{1}{2} = .5$.

22. **The correct answer is (D).** Using log rules we can first rewrite the original equation to $\log_2 (x - 3)(x - 2) = 1$. Converting this log equation to exponential form, we have $(x - 3)(x - 2) = 2^1$ or $x^2 - 5x + 4 = 0$. This factors to $(x - 4)(x - 1) = 0$ and so $x = 4$ or $x = 1$. However we see that if $x = 1$ then $x - 3 < 0$ and we can not take logs of negative numbers. So only $x = 4$ works.

23. **The correct answer is (D).** In the worst case, a person would select a different towel on the first three selections. We see, though, on his or her fourth selection, the towel would match one of the previous three.

24. **The correct answer is (C).** Look at the pattern. The second number is just the first number plus 6. The third number is just the second number plus 2 times 6. The fourth number is just the third number plus 3 times 6 and so on. So, the next number in the sequence is the previous number plus 5 times 6 or 91.

25. **The correct answer is (B).** For the two squares, there are eight triangles for each square and two in the intersection, for a total of eighteen triangles. For the three squares, there are eight triangles for each square; for each of the two pairs of squares that intersect, there are two squares; and for all three squares together, there is one more triangle for a total of $8 + 8 + 8 + 2 + 2 + 1 = 29$ squares.

26. **The correct answer is (D).** $f(x) = x^2$ is a polynomial (all coefficients are 0 except $a_2 = 1$). It is even since $f(-x) = f(x)$, and it is a quadratic. (It has the form of $ax^2 + bx^2 + c$).

27. **The correct answer is (C).** $\tan(-\eta)\cos(-\eta) = -\tan\eta \cos\eta = -\sin\eta$.

28. **The correct answer is (D).** By definition, you mirror about $y = x$.

29. **The correct answer is (C).** For pages 1 through 9, there are 9 digits used, for pages 10 through 99, there are $2(90) = 180$ digits used, and for pages 100 through 999, there are $3(900) = 2,700$ digits used. So, there are 568 digits left for pages 1,000 through 1,141. The book has 1,141 pages.

30. **The correct answer is (C).**

$$\left(1 - \frac{x}{y}\right) \div \left(1 - \frac{x^2}{y}\right) = \left(\frac{y-x}{y}\right) \div \left(\frac{y^2 - x^2}{y}\right)$$

$$= \left(\frac{y-x}{y}\right) \times \left(\frac{y}{y^2 - x^2}\right)$$

$$= \frac{1}{y+x} = (y+x)^{-1}.$$

31. **The correct answer is (C).** Using the Law of Cosines,

    $c^2 = a^2 + b^2 - 2ab\cos C$
    $= (4.21)^2 + (5.71)^2 - 2(4.21)(5.71)\cos(28.3)$
    $= 7.996433703 \approx 8.00$.

    So $c = \sqrt{8} = 2.828$.

32. **The correct answer is (B).** Using properties of similar triangles, we have the ratio $\frac{12 + x}{12} = \frac{30}{20}$, which implies that $x = 6$.

33. **The correct answer is (A).** $(2x + 1)^{x^2 - 4} = 1$ when $x^2 - 4 = 0$ or when $2x + 1 = 1$ since anything except 0 to the 0th power is 1 and 1 to any power is 1. So, $x = 2$ and $x = 0$.

34. **The correct answer is (B).** Work out some simple examples to see the total number of subsets for a set with m elements. If the set has just one element then the subsets are the empty set and the set itself. Thus, there are $2^1 = 2$ subsets of a one-element set. If the set has two elements—say the set is {1,2}—then the subsets are φ, {1}, {2}, {1, 2}. Thus, there are $2^2 = 4$ subsets of a set of two elements. If the set has three elements—say the set is {1, 2, 3}—then the subsets are φ, {1}, {2}, {3}, {1, 2}, {1, 3}, {2, 3}, {1, 2, 3}. Thus, the set has $2^3 = 8$ subsets. Continue this argument and one sees that for a set {1, 2, 3, 4, ..., m − 1, m}, there are 2^m subsets.

35. **The correct answer is (C).** $V = \frac{((5)(12)(20))}{2} = 600$ cubed centimeters.

36. **The correct answer is (B).** The axis of symmetry is $x = \frac{-b}{2a} = \frac{-(-2)}{2(1)} = 1$.

37. **The correct answer is (E).** The numbers are 2, 3, 4, 5, and 6 and either 1 or 7. So, the possible products are 7! or 6!. Choice (E) is the correct answer.

38. **The correct answer is (A).** We see that $ax^n - bx = x(ax^{n-1} - b) = 0$. Since x is not equal to 0, we have $b = ax^{n-1}$.

ANSWERS AND EXPLANATIONS

39. **The correct answer is (D).** We have
$$f\left(\frac{\pi}{2}\right) = 2\sin\left(3\left(\frac{\pi}{2}\right)\right)$$
$$= 2\sin\left(\frac{3\pi}{2}\right)$$
$$= 2(-1)$$
$$= -2.$$

40. **The correct answer is (B).** Use the identity for $\cos 2x$ to get
$$\sin^2 x + \frac{(\cos^2 x - \sin^2 x)}{2} + \frac{1}{2} = \frac{(\sin^2 x)}{2} + \frac{(\cos^2 x)}{2} + \frac{1}{2} = 1.$$
The last equality holds because $\sin^2 x + \cos^2 x = 1$.

41. **The correct answer is (B).** If $C < 0$, then $-C > 0$ and $4 - C > 4 > 0$. Also, $4 - C$ is not equal to $-C$ because adding C to both sides gives the incorrect equation, $4 = 0$. So, the expressions in the denominator are different and, since we can rewrite the equation to $\frac{x^2}{4-C} + \frac{y^2}{-C} = 1$ and the expressions in the denominator are both positive, we know the equation is an equation for an ellipse.

42. **The correct answer is (A).** Look at the pattern of powers of 3.

$3^0 = 1 \quad 3^4 = 81 \quad 3^8 = 6,561$
$3^1 = 3 \quad 3^5 = 243 \quad \text{etc.}$
$3^2 = 9 \quad 3^6 = 729$

We notice the there is a repeated pattern of 1, 3, 9, and 7 for the units' digits of powers on 3. So, we can calculate the units digit of 3^{30} by dividing 30 by 4 and seeing the remainder 2. The units digit of 3^{30} is equal to the units digit of 3^2, which we know to be 9.

43. **The correct answer is (A).** Let A_1 be the area of the 8-by-6 rectangle and A_2 be the area of the 4 by 3 rectangle. Also, let A be the area of the overlap. So, the difference in the non overlapping regions is
$A_d = (A_1 - A) - (A_2 - A) = A_1 - A_2 = 48 - 12 = 36.$

44. **The correct answer is (B).** If $x^{-2} = 36$, then $x^2 = \frac{1}{36}$. So, $x = \pm\frac{1}{6}$.

45. **The correct answer is (E).** Draw a picture of the ball room and calculate d.

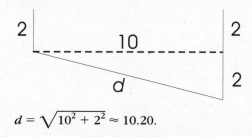

$d = \sqrt{10^2 + 2^2} \approx 10.20.$

PRACTICE TEST 2

46. **The correct answer is (A).** The two equations in standard form are $y = 3x - 9$ and $y = \frac{-k}{3}x + \frac{5}{3}$. So, using properties of perpendicular slopes, we have that $\frac{-k}{3} = \frac{-1}{3}$ or $k = 1$.

47. **The correct answer is (B).** This is an arithmetic sequence. 1, 3, 5, ..., 81. We know the sum of such a sequence is

$$S = \frac{n}{2} \cdot [2a + (n-1)d]$$

$$= \frac{40}{2} \cdot [2 \cdot 1 + 39 \cdot 2]$$

$$= 20 \cdot 80 = 1,600.$$

Therefore, the sum of the first 40 odd numbers is 1,600.

48. **The correct answer is (C).** We know $f(g(1)) = 5$, which implies that $2(g(1)) + 3 = 5$ or $g(1) = 1$. The only equation that this condition holds is choice (C).

49. **The correct answer is (C).** Set up the equation $\frac{1}{4} + \frac{1}{3} = \frac{1}{x}$. The one fourth stands for Bill completing 1 house in 4 hours, and the one third stands for John completing 1 house in 3 hours. Combine these together to see how long x it takes to complete 1 house. Solving for x, we get $x = \frac{4(3)}{4+3} \approx 1.71$.

50. **The correct answer is (C).** Set up the equation $\cos 45 = \frac{4}{AC}$ or $AC \frac{4}{\cos 45} = 4\sqrt{2}$. So, the perimeter is $P = 2(6) + 2(4\sqrt{2}) \approx 23.31$.

Answer Sheet

Diagnostic 1
Mathematics Level IC

1 Ⓐ Ⓑ Ⓒ Ⓓ Ⓔ	8 Ⓐ Ⓑ Ⓒ Ⓓ Ⓔ	15 Ⓐ Ⓑ Ⓒ Ⓓ Ⓔ	22 Ⓐ Ⓑ Ⓒ Ⓓ Ⓔ
2 Ⓐ Ⓑ Ⓒ Ⓓ Ⓔ	9 Ⓐ Ⓑ Ⓒ Ⓓ Ⓔ	16 Ⓐ Ⓑ Ⓒ Ⓓ Ⓔ	23 Ⓐ Ⓑ Ⓒ Ⓓ Ⓔ
3 Ⓐ Ⓑ Ⓒ Ⓓ Ⓔ	10 Ⓐ Ⓑ Ⓒ Ⓓ Ⓔ	17 Ⓐ Ⓑ Ⓒ Ⓓ Ⓔ	24 Ⓐ Ⓑ Ⓒ Ⓓ Ⓔ
4 Ⓐ Ⓑ Ⓒ Ⓓ Ⓔ	11 Ⓐ Ⓑ Ⓒ Ⓓ Ⓔ	18 Ⓐ Ⓑ Ⓒ Ⓓ Ⓔ	25 Ⓐ Ⓑ Ⓒ Ⓓ Ⓔ
5 Ⓐ Ⓑ Ⓒ Ⓓ Ⓔ	12 Ⓐ Ⓑ Ⓒ Ⓓ Ⓔ	19 Ⓐ Ⓑ Ⓒ Ⓓ Ⓔ	
6 Ⓐ Ⓑ Ⓒ Ⓓ Ⓔ	13 Ⓐ Ⓑ Ⓒ Ⓓ Ⓔ	20 Ⓐ Ⓑ Ⓒ Ⓓ Ⓔ	
7 Ⓐ Ⓑ Ⓒ Ⓓ Ⓔ	14 Ⓐ Ⓑ Ⓒ Ⓓ Ⓔ	21 Ⓐ Ⓑ Ⓒ Ⓓ Ⓔ	

Diagnostic 2
Mathematics Level IIc

1 Ⓐ Ⓑ Ⓒ Ⓓ Ⓔ	8 Ⓐ Ⓑ Ⓒ Ⓓ Ⓔ	15 Ⓐ Ⓑ Ⓒ Ⓓ Ⓔ	22 Ⓐ Ⓑ Ⓒ Ⓓ Ⓔ
2 Ⓐ Ⓑ Ⓒ Ⓓ Ⓔ	9 Ⓐ Ⓑ Ⓒ Ⓓ Ⓔ	16 Ⓐ Ⓑ Ⓒ Ⓓ Ⓔ	23 Ⓐ Ⓑ Ⓒ Ⓓ Ⓔ
3 Ⓐ Ⓑ Ⓒ Ⓓ Ⓔ	10 Ⓐ Ⓑ Ⓒ Ⓓ Ⓔ	17 Ⓐ Ⓑ Ⓒ Ⓓ Ⓔ	24 Ⓐ Ⓑ Ⓒ Ⓓ Ⓔ
4 Ⓐ Ⓑ Ⓒ Ⓓ Ⓔ	11 Ⓐ Ⓑ Ⓒ Ⓓ Ⓔ	18 Ⓐ Ⓑ Ⓒ Ⓓ Ⓔ	25 Ⓐ Ⓑ Ⓒ Ⓓ Ⓔ
5 Ⓐ Ⓑ Ⓒ Ⓓ Ⓔ	12 Ⓐ Ⓑ Ⓒ Ⓓ Ⓔ	19 Ⓐ Ⓑ Ⓒ Ⓓ Ⓔ	
6 Ⓐ Ⓑ Ⓒ Ⓓ Ⓔ	13 Ⓐ Ⓑ Ⓒ Ⓓ Ⓔ	20 Ⓐ Ⓑ Ⓒ Ⓓ Ⓔ	
7 Ⓐ Ⓑ Ⓒ Ⓓ Ⓔ	14 Ⓐ Ⓑ Ⓒ Ⓓ Ⓔ	21 Ⓐ Ⓑ Ⓒ Ⓓ Ⓔ	

Practice Test 1
Mathematics Level IC

1 Ⓐ Ⓑ Ⓒ Ⓓ Ⓔ	16 Ⓐ Ⓑ Ⓒ Ⓓ Ⓔ	31 Ⓐ Ⓑ Ⓒ Ⓓ Ⓔ	46 Ⓐ Ⓑ Ⓒ Ⓓ Ⓔ
2 Ⓐ Ⓑ Ⓒ Ⓓ Ⓔ	17 Ⓐ Ⓑ Ⓒ Ⓓ Ⓔ	32 Ⓐ Ⓑ Ⓒ Ⓓ Ⓔ	47 Ⓐ Ⓑ Ⓒ Ⓓ Ⓔ
3 Ⓐ Ⓑ Ⓒ Ⓓ Ⓔ	18 Ⓐ Ⓑ Ⓒ Ⓓ Ⓔ	33 Ⓐ Ⓑ Ⓒ Ⓓ Ⓔ	48 Ⓐ Ⓑ Ⓒ Ⓓ Ⓔ
4 Ⓐ Ⓑ Ⓒ Ⓓ Ⓔ	19 Ⓐ Ⓑ Ⓒ Ⓓ Ⓔ	34 Ⓐ Ⓑ Ⓒ Ⓓ Ⓔ	49 Ⓐ Ⓑ Ⓒ Ⓓ Ⓔ
5 Ⓐ Ⓑ Ⓒ Ⓓ Ⓔ	20 Ⓐ Ⓑ Ⓒ Ⓓ Ⓔ	35 Ⓐ Ⓑ Ⓒ Ⓓ Ⓔ	50 Ⓐ Ⓑ Ⓒ Ⓓ Ⓔ
6 Ⓐ Ⓑ Ⓒ Ⓓ Ⓔ	21 Ⓐ Ⓑ Ⓒ Ⓓ Ⓔ	36 Ⓐ Ⓑ Ⓒ Ⓓ Ⓔ	
7 Ⓐ Ⓑ Ⓒ Ⓓ Ⓔ	22 Ⓐ Ⓑ Ⓒ Ⓓ Ⓔ	37 Ⓐ Ⓑ Ⓒ Ⓓ Ⓔ	
8 Ⓐ Ⓑ Ⓒ Ⓓ Ⓔ	23 Ⓐ Ⓑ Ⓒ Ⓓ Ⓔ	38 Ⓐ Ⓑ Ⓒ Ⓓ Ⓔ	
9 Ⓐ Ⓑ Ⓒ Ⓓ Ⓔ	24 Ⓐ Ⓑ Ⓒ Ⓓ Ⓔ	39 Ⓐ Ⓑ Ⓒ Ⓓ Ⓔ	
10 Ⓐ Ⓑ Ⓒ Ⓓ Ⓔ	25 Ⓐ Ⓑ Ⓒ Ⓓ Ⓔ	40 Ⓐ Ⓑ Ⓒ Ⓓ Ⓔ	
11 Ⓐ Ⓑ Ⓒ Ⓓ Ⓔ	26 Ⓐ Ⓑ Ⓒ Ⓓ Ⓔ	41 Ⓐ Ⓑ Ⓒ Ⓓ Ⓔ	
12 Ⓐ Ⓑ Ⓒ Ⓓ Ⓔ	27 Ⓐ Ⓑ Ⓒ Ⓓ Ⓔ	42 Ⓐ Ⓑ Ⓒ Ⓓ Ⓔ	
13 Ⓐ Ⓑ Ⓒ Ⓓ Ⓔ	28 Ⓐ Ⓑ Ⓒ Ⓓ Ⓔ	43 Ⓐ Ⓑ Ⓒ Ⓓ Ⓔ	
14 Ⓐ Ⓑ Ⓒ Ⓓ Ⓔ	29 Ⓐ Ⓑ Ⓒ Ⓓ Ⓔ	44 Ⓐ Ⓑ Ⓒ Ⓓ Ⓔ	
15 Ⓐ Ⓑ Ⓒ Ⓓ Ⓔ	30 Ⓐ Ⓑ Ⓒ Ⓓ Ⓔ	45 Ⓐ Ⓑ Ⓒ Ⓓ Ⓔ	

Practice Test 2

Mathematics Level IC

1	A B C D E	16	A B C D E	31	A B C D E	46	A B C D E
2	A B C D E	17	A B C D E	32	A B C D E	47	A B C D E
3	A B C D E	18	A B C D E	33	A B C D E	48	A B C D E
4	A B C D E	19	A B C D E	34	A B C D E	49	A B C D E
5	A B C D E	20	A B C D E	35	A B C D E	50	A B C D E
6	A B C D E	21	A B C D E	36	A B C D E		
7	A B C D E	22	A B C D E	37	A B C D E		
8	A B C D E	23	A B C D E	38	A B C D E		
9	A B C D E	24	A B C D E	39	A B C D E		
10	A B C D E	25	A B C D E	40	A B C D E		
11	A B C D E	26	A B C D E	41	A B C D E		
12	A B C D E	27	A B C D E	42	A B C D E		
13	A B C D E	28	A B C D E	43	A B C D E		
14	A B C D E	29	A B C D E	44	A B C D E		
15	A B C D E	30	A B C D E	45	A B C D E		

Practice Test 1

Mathematics Level IIC

1	A B C D E	16	A B C D E	31	A B C D E	46	A B C D E
2	A B C D E	17	A B C D E	32	A B C D E	47	A B C D E
3	A B C D E	18	A B C D E	33	A B C D E	48	A B C D E
4	A B C D E	19	A B C D E	34	A B C D E	49	A B C D E
5	A B C D E	20	A B C D E	35	A B C D E	50	A B C D E
6	A B C D E	21	A B C D E	36	A B C D E		
7	A B C D E	22	A B C D E	37	A B C D E		
8	A B C D E	23	A B C D E	38	A B C D E		
9	A B C D E	24	A B C D E	39	A B C D E		
10	A B C D E	25	A B C D E	40	A B C D E		
11	A B C D E	26	A B C D E	41	A B C D E		
12	A B C D E	27	A B C D E	42	A B C D E		
13	A B C D E	28	A B C D E	43	A B C D E		
14	A B C D E	29	A B C D E	44	A B C D E		
15	A B C D E	30	A B C D E	45	A B C D E		

Practice Test 2

Mathematics Level IIC

1	A B C D E	16	A B C D E	31	A B C D E	46	A B C D E
2	A B C D E	17	A B C D E	32	A B C D E	47	A B C D E
3	A B C D E	18	A B C D E	33	A B C D E	48	A B C D E
4	A B C D E	19	A B C D E	34	A B C D E	49	A B C D E
5	A B C D E	20	A B C D E	35	A B C D E	50	A B C D E
6	A B C D E	21	A B C D E	36	A B C D E		
7	A B C D E	22	A B C D E	37	A B C D E		
8	A B C D E	23	A B C D E	38	A B C D E		
9	A B C D E	24	A B C D E	39	A B C D E		
10	A B C D E	25	A B C D E	40	A B C D E		
11	A B C D E	26	A B C D E	41	A B C D E		
12	A B C D E	27	A B C D E	42	A B C D E		
13	A B C D E	28	A B C D E	43	A B C D E		
14	A B C D E	29	A B C D E	44	A B C D E		
15	A B C D E	30	A B C D E	45	A B C D E		